FREE Test Taking Tips DVD Offer

To help us better serve you, we have developed a Test Taking Tips DVD that we would like to give you for FREE. **This DVD covers world-class test taking tips that you can use to be even more successful when you are taking your test.**

All that we ask is that you email us your feedback about your study guide. Please let us know what you thought about it – whether that is good, bad or indifferent.

To get your **FREE Test Taking Tips DVD**, email freedvd@studyguideteam.com with "FREE DVD" in the subject line and the following information in the body of the email:

a. The title of your study guide.

b. Your product rating on a scale of 1-5, with 5 being the highest rating.

c. Your feedback about the study guide. What did you think of it?

d. Your full name and shipping address to send your free DVD.

If you have any questions or concerns, please don't hesitate to contact us at freedvd@studyguideteam.com.

Thanks again!

ACCUPLACER Math Study Guide

ACCUPLACER Math Test Prep Team

Table of Contents

Quick Overview

As you draw closer to taking your exam, effective preparation becomes more and more important. Thankfully, you have this study guide to help you get ready. Use this guide to help keep your studying on track and refer to it often.

This study guide contains several key sections that will help you be successful on your exam. The guide contains tips for what you should do the night before and the day of the test. Also included are test-taking tips. Knowing the right information is not always enough. Many well-prepared test takers struggle with exams. These tips will help equip you to accurately read, assess, and answer test questions.

A large part of the guide is devoted to showing you what content to expect on the exam and to helping you better understand that content. Near the end of this guide is a practice test so that you can see how well you have grasped the content. Then, answers explanations are provided so that you can understand why you missed certain questions.

Don't try to cram the night before you take your exam. This is not a wise strategy for a few reasons. First, your retention of the information will be low. Your time would be better used by reviewing information you already know rather than trying to learn a lot of new information. Second, you will likely become stressed as you try to gain large amount of knowledge in a short amount of time. Third, you will be depriving yourself of sleep. So be sure to go to bed at a reasonable time the night before. Being well-rested helps you focus and remain calm.

Be sure to eat a substantial breakfast the morning of the exam. If you are taking the exam in the afternoon, be sure to have a good lunch as well. Being hungry is distracting and can make it difficult to focus. You have hopefully spent lots of time preparing for the exam. Don't let an empty stomach get in the way of success!

When travelling to the testing center, leave earlier than needed. That way, you have a buffer in case you experience any delays. This will help you remain calm and will keep you from missing your appointment time at the testing center.

Be sure to pace yourself during the exam. Don't try to rush through the exam. There is no need to risk performing poorly on the exam just so you can leave the testing center early. Allow yourself to use all of the allotted time if needed.

Remain positive while taking the exam even if you feel like you are performing poorly. Thinking about the content you should have mastered will not help you perform better on the exam.

Once the exam is complete, take some time to relax. Even if you feel that you need to take the exam again, you will be well served by some down time before you begin studying again. It's often easier to convince yourself to study if you know that it will come with a reward!

Test-Taking Strategies

1. Predicting the Answer

When you feel confident in your preparation for a multiple-choice test, try predicting the answer before reading the answer choices. This is especially useful on questions that test objective factual knowledge or that ask you to fill in a blank. By predicting the answer before reading the available choices, you eliminate the possibility that you will be distracted or led astray by an incorrect answer choice. You will feel more confident in your selection if you read the question, predict the answer, and then find your prediction among the answer choices. After using this strategy, be sure to still read all of the answer choices carefully and completely. If you feel unprepared, you should not attempt to predict the answers. This would be a waste of time and an opportunity for your mind to wander in the wrong direction.

2. Reading the Whole Question

Too often, test takers scan a multiple-choice question, recognize a few familiar words, and immediately jump to the answer choices. Test authors are aware of this common impatience, and they will sometimes prey upon it. For instance, a test author might subtly turn the question into a negative, or he or she might redirect the focus of the question right at the end. The only way to avoid falling into these traps is to read the entirety of the question carefully before reading the answer choices.

3. Looking for Wrong Answers

Long and complicated multiple-choice questions can be intimidating. One way to simplify a difficult multiple-choice question is to eliminate all of the answer choices that are clearly wrong. In most sets of answers, there will be at least one selection that can be dismissed right away. If the test is administered on paper, the test taker could draw a line through it to indicate that it may be ignored; otherwise, the test taker will have to perform this operation mentally or on scratch paper. In either case, once the obviously incorrect answers have been eliminated, the remaining choices may be considered. Sometimes identifying the clearly wrong answers will give the test taker some information about the correct answer. For instance, if one of the remaining answer choices is a direct opposite of one of the eliminated answer choices, it may well be the correct answer. The opposite of obviously wrong is obviously right! Of course, this is not always the case. Some answers are obviously incorrect simply because they are irrelevant to the question being asked. Still, identifying and eliminating some incorrect answer choices is a good way to simplify a multiple-choice question.

4. Don't Overanalyze

Anxious test takers often overanalyze questions. When you are nervous, your brain will often run wild causing you to make associations and discover clues that don't actually exist. If you feel that this may be a problem for you, do whatever you can to slow down during the test. Try taking a deep breath or counting to ten. As you read and consider the question, restrict yourself to the particular words used by the author. Avoid thought tangents about what the author *really* meant, or what he or she was *trying* to say. The only things that matter on a multiple-choice test are the words that are actually in the question. You must avoid reading too much into a multiple-choice question, or supposing that the writer meant something other than what he or she wrote.

5. No Need for Panic

It is wise to learn as many strategies as possible before taking a multiple-choice test, but it is likely that you will come across a few questions for which you simply don't know the answer. In this situation, avoid panicking. Because most multiple-choice tests include dozens of questions, the relative value of a single wrong answer is small. Moreover, your failure on one question has no effect on your success elsewhere on the test. As much as possible, you should compartmentalize each question on a multiple-choice test. In other words, you should not allow your feelings about one question to affect your success on the others. When you find a question that you either don't understand or don't know how to answer, just take a deep breath and do your best. Read the entire question slowly and carefully. Try rephrasing the question a couple of different ways. Then, read all of the answer choices carefully. After eliminating obviously wrong answers, make a selection and move on to the next question.

6. Confusing Answer Choices

When working on a difficult multiple-choice question, there may be a tendency to focus on the answer choices that are the easiest to understand. Many people, whether consciously or not, gravitate to the answer choices that require the least concentration, knowledge, and memory. This is a mistake. When you come across an answer choice that is confusing, you need to give it extra attention. A question might be confusing because you do not know the subject matter to which it refers. If this is the case, don't eliminate the answer before you have affirmatively settled on another. When you come across an answer choice of this type, set it aside as you look at the remaining choices. If you can confidently assert that one of the other choices is correct, you can leave the confusing answer aside. Otherwise, you will need to take a moment to try to better understand the confusing answer choice. Rephrasing is one way to tease out the sense of a confusing answer choice.

7. Your First Instinct

Many people struggle with multiple-choice tests because they overthink the questions. If you have studied sufficiently for the test, you should be prepared to trust your first instinct once you have carefully and completely read the question and all of the answer choices. There is a great deal of research suggesting that the mind can come to the correct conclusion very quickly once it has obtained all of the relevant information. At times, it may seem to you as if your intuition is working faster even than your reasoning mind. This may in fact be true. The knowledge you obtain while studying may be retrieved from your subconscious before you have a chance to work out the associations that support it. Verify your instinct by working out the reasons that it should be trusted.

8. Key Words

Many test takers struggle with multiple-choice questions because they have poor reading comprehension skills. Quickly reading and understanding a multiple-choice question requires a mixture of skill and experience. To help with this, try jotting down a few key words and phrases on a piece of scrap paper. Doing this concentrates the process of reading and forces the mind to weigh the relative importance of the question's parts. In selecting words and phrases to write down, the test taker thinks about the question more deeply and carefully. This is especially true for multiple-choice questions that are preceded by a long prompt.

9. Subtle Negatives

One of the oldest tricks in the multiple-choice test writer's book is to subtly reverse the meaning of a question with a word like *not* or *except*. If you are not paying attention to each word in the question, you can easily be led astray by this trick. For instance, a common question format is, "Which of the following is...?" Obviously, if the question instead is, "Which of the following is not....?," then the answer will be quite different. Even worse, the test makers are aware of the potential for this mistake and will include one answer choice that would be correct if the question were not negated or reversed. A test taker who misses the reversal will find what he or she believes to be a correct answer and will be so confident that he or she will fail to reread the question and discover the original error. The only way to avoid this is to practice a wide variety of multiple-choice questions and to pay close attention to each and every word.

10. Reading Every Answer Choice

It may seem obvious, but you should always read every one of the answer choices! Too many test takers fall into the habit of scanning the question and assuming that they understand the question because they recognize a few key words. From there, they pick the first answer choice that answers the question they believe they have read. Test takers who read all of the answer choices might discover that one of the latter answer choices is actually *more* correct. Moreover, reading all of the answer choices can remind you of facts related to the question that can help you arrive at the correct answer. Sometimes, a misstatement or incorrect detail in one of the latter answer choices will trigger your memory of the subject and will enable you to find the right answer. Failing to read all of the answer choices is like not reading all of the items on a restaurant menu: you might miss out on the perfect choice.

11. Spot the Hedges

One of the keys to success on multiple-choice tests is paying close attention to every word. This is never more true than with words like *almost*, *most*, *some*, and *sometimes*. These words are called "hedges", because they indicate that a statement is not totally true or not true in every place and time. An absolute statement will contain no hedges, but in many subjects, like literature and history, the answers are not always straightforward or absolute. There are always exceptions to the rules in these subjects. For this reason, you should favor those multiple-choice questions that contain hedging language. The presence of qualifying words indicates that the author is taking special care with his or her words, which is certainly important when composing the right answer. After all, there are many ways to be wrong, but there is only one way to be right! For this reason, it is wise to avoid answers that are absolute when taking a multiple-choice test. An absolute answer is one that says things are either all one way or all another. They often include words like *every*, *always*, *best*, and *never*. If you are taking a multiple-choice test in a subject that doesn't lend itself to absolute answers, be on your guard if you see any of these words.

12. Long Answers

In many subject areas, the answers are not simple. As already mentioned, the right answer often requires hedges. Another common feature of the answers to a complex or subjective question are qualifying clauses, which are groups of words that subtly modify the meaning of the sentence. If the question or answer choice describes a rule to which there are exceptions or the subject matter is complicated, ambiguous, or confusing, the correct answer will require many words in order to be expressed clearly and accurately. In essence, you should not be deterred by answer choices that seem excessively long. Oftentimes, the author of the text will not be able to write the correct answer without offering some qualifications and modifications. Your job is to read the answer choices thoroughly and completely and to select the one that most accurately and precisely answers the question.

13. Restating to Understand

Sometimes, a question on a multiple-choice test is difficult not because of what it asks but because of how it is written. If this is the case, restate the question or answer choice in different words. This process serves a couple of important purposes. First, it forces you to concentrate on the core of the question. In order to rephrase the question accurately, you have to understand it well. Rephrasing the question will concentrate your mind on the key words and ideas. Second, it will present the information to your mind in a fresh way. This process may trigger your memory and render some useful scrap of information picked up while studying.

14. True Statements

Sometimes an answer choice will be true in itself, but it does not answer the question. This is one of the main reasons why it is essential to read the question carefully and completely before proceeding to the answer choices. Too often, test takers skip ahead to the answer choices and look for true statements. Having found one of these, they are content to select it without reference to the question above. Obviously, this provides an easy way for test makers to play tricks. The savvy test taker will always read the entire question before turning to the answer choices. Then, having settled on a correct answer choice, he or she will refer to the original question and ensure that the selected answer is relevant. The mistake of choosing a correct-but-irrelevant answer choice is especially common on questions related to specific pieces of objective knowledge, like historical or scientific facts. A prepared test taker will have a wealth of factual knowledge at his or her disposal, and should not be careless in its application.

15. No Patterns

One of the more dangerous ideas that circulates about multiple-choice tests is that the correct answers tend to fall into patterns. These erroneous ideas range from a belief that B and C are the most common right answers, to the idea that an unprepared test-taker should answer "A-B-A-C-A-D-A-B-A." It cannot be emphasized enough that pattern-seeking of this type is exactly the WRONG way to approach a multiple-choice test. To begin with, it is highly unlikely that the test maker will plot the correct answers according to some predetermined pattern. The questions are scrambled and delivered in a random order. Furthermore, even if the test maker was following a pattern in the assignation of correct answers, there is no reason why the test taker would know which pattern he or she was using. Any attempt to discern a pattern in the answer choices is a waste of time and a distraction from the real work of taking the test. A test taker would be much better served by extra preparation before the test than by reliance on a pattern in the answers.

FREE DVD OFFER

Don't forget that doing well on your exam includes both understanding the test content and understanding how to use what you know to do well on the test. We offer a completely FREE Test Taking Tips DVD that covers world class test taking tips that you can use to be even more successful when you are taking your test.

All that we ask is that you email us your feedback about your study guide. To get your **FREE Test Taking Tips DVD**, email freedvd@studyguideteam.com with "FREE DVD" in the subject line and the following information in the body of the email:

- The title of your study guide.
- Your product rating on a scale of 1-5, with 5 being the highest rating.
- Your feedback about the study guide. What did you think of it?
- Your full name and shipping address to send your free DVD.

Introduction to the ACCUPLACER

Function of the Test

ACCUPLACER is an adaptive, computerized test offered by the College Board and used by some colleges and high schools to determine placement of students in programs appropriate to the students' skill level. The test is offered nation-wide, at any college or high school that chooses to use it. Test-takers are almost always students of schools that use the ACCUPLACER in their course selection and placement efforts. Schools can also use the ACCUPLACER to assess students' skill levels and identify specific areas in which the students need improvement. Scores are generally used only by the college or high school the student is already attending for placement and instruction at that school.

According to the College Board, more than 1,500 secondary and post-secondary institutions are currently using the ACCUPLACER. Over 7.5 million ACCUPLACER tests are administered in a typical year. The College Board recommends that ACCUPLACER scores be used in conjunction with other variables including high school GPA, the number of years a student has taken coursework in a particular subject area, other test scores such as the SAT or ACT, as well as non-cognitive information such as motivation, family support, and time management skills, in order to place students in courses of appropriate difficulty.

Test Administration

ACCUPLACER is offered by computer, usually by the school that wants to use its results, whenever a student makes an appointment with the school to take it. In cases where a student is not able to take the test at the student's school, arrangements can sometimes be made to take the test at a more convenient location on another school's campus.

Fees for taking the ACCUPLACER vary from school to school. Students sometimes do not pay a fee at all; rather, the school pays the College Board for the right to administer the test and then students take it for free. Students may retake the ACCUPLACER at the discretion of the school administering the test and using its results. Students with documented disabilities can make arrangements, through the test center offering the ACCUPLACER, to take the test with appropriate accommodations, including the potential availability of a written version of the test.

Test Format

The core of the ACCUPLACER is five multiple-choice subject area tests: Arithmetic, College-level Math, Elementary Algebra, Reading Comprehension, and Sentence Skills. A college may ask a student to test in any or all of the five subject areas, depending on the student's and the school's needs. There is also a sixth section, called the WritePlacer, in which students must write a brief essay. Finally, there are four multiple-choice ESL sections of the ACCUPLACER, which schools may ask students, for whom English is a second language, to take.

The multiple-choice sections of the ACCUPLACER are administered by computer. There is no time-limit, so test takers should feel free to work at their own pace. A typical test section might take around 30 minutes to complete. The test is adaptive, meaning it adjusts the difficulty of each question based on the student's success on the previous questions. The more questions a test-taker gets right, the harder succeeding questions will be and vice-versa. This allows the test to more readily determine test-takers' ability levels without wasting time on questions far above or far below what the test-takers can answer.

Section	Questions	Description
Arithmetic	17	Basic arithmetic and problem solving
College Level Math	20	Algebra, geometry, trigonometry
Elementary Algebra	12	Basic algebra
Reading Comprehension	20	Understanding reading; making inferences
Sentence Skills	20	Sentence structure; relationships between sentences
WritePlacer	1 essay	Effective written communication
ESL- Language Use	20	English grammar
ESL- Listening	20	Understanding spoken English communication
ESL- Reading Skills	20	Comprehension of short written English passages
ESL- Sentence Meaning	20	Understanding the meaning of English sentences

Scoring

The ACCUPLACER subject area tests are scored on a scale from 20 to 120. Schools are free to use the scores for placement as they see fit, given that the difficulty of coursework varies from school to school. A typical community college might separate scores into tiered groups from 50 to 75, from 75 to 99, and from 100 to 120. Scores are typically generated instantly by the computer, but may not be available from the school administering the test until the school has had a chance to review the scores and use them for placement purposes.

Recent/Future Developments

Starting in September 2016, the College Board will be making "next-generation" ACCUPLACER tests available to schools. Schools will have the option of administering the next-generation tests or the older tests, but not both. The new next-generation tests will include redesigned reading, writing, and math content, intended to more effectively help schools place students in classes that match their skill level.

Arithmetic

Operations with Whole Numbers and Fractions

Addition with Whole Numbers and Fractions

Addition combines two quantities together. With whole numbers, this is taking two sets of things and merging them into one, then counting the result. For example, 4 + 3 = 7. When adding numbers, the order does not matter: 3 + 4 = 7, also. Longer lists of whole numbers can also be added together. The result of adding numbers is called the *sum*.

With fractions, the number on top is the *numerator*, and the number on the bottom is the *denominator*. To add fractions, the denominator must be the same—a *common denominator*. To find a common denominator, the existing numbers on the bottom must be considered, and the lowest number they will both multiply into must be determined. Consider the following equation:

$$\frac{1}{3}+\frac{5}{6}=?$$

The numbers 3 and 6 both multiply into 6. Three can be multiplied by 2, and 6 can be multiplied by 1. The top and bottom of each fraction must be multiplied by the same number. Then, the numerators are added together to get a new numerator. The following equation is the result:

$$\frac{1}{3}+\frac{5}{6}=\frac{2}{6}+\frac{5}{6}=\frac{7}{6}$$

Subtraction with Whole Numbers and Fractions

Subtraction is taking one quantity away from another, so it is the opposite of addition. The expression 4 – 3 means taking 3 away from 4. So, 4 – 3 = 1. In this case, the order matters, since it entails taking one quantity away from the other, rather than just putting two quantities together. The result of subtraction is also called the *difference*.

To subtract fractions, the denominator must be the same. Then, subtract the numerators together to get a new numerator. Here is an example:

$$\frac{1}{3}-\frac{5}{6}=\frac{2}{6}-\frac{5}{6}=\frac{-3}{6}=-\frac{1}{2}$$

Multiplication with Whole Numbers and Fractions

Multiplication is a kind of repeated addition. The expression 4×5 is taking four sets, each of them having five things in them, and putting them all together. That means $4 \times 5 = 5 + 5 + 5 + 5 = 20$. As with addition, the order of the numbers does not matter. The result of a multiplication problem is called the *product.*

To multiply fractions, the numerators are multiplied to get the new numerator, and the denominators are multiplied to get the new denominator:

$$\frac{1}{3}\times\frac{5}{6}=\frac{1\times 5}{3\times 6}=\frac{5}{18}$$

When multiplying fractions, common factors can *cancel* or *divide into one another*, when factors that appear in the numerator of one fraction and the denominator of the other fraction. Here is an example:

$$\frac{1}{3} \times \frac{9}{8} = \frac{1}{1} \times \frac{3}{8} = 1 \times \frac{3}{8} = \frac{3}{8}$$

The numbers 3 and 9 have a common factor of 3, so that factor can be divided out.

Division with Whole Numbers and Fractions

Division is the opposite of multiplication. With whole numbers, it means splitting up one number into sets of equal size. For example, $16 \div 8$ is the number of sets of eight things that can be made out of sixteen things. Thus, $16 \div 8 = 2$. As with subtraction, the order of the numbers will make a difference, here. The answer to a division problem is called the *quotient*, while the number in front of the division sign is called the *dividend* and the number behind the division sign is called the *divisor*.

To divide fractions, the first fraction must be multiplied with the reciprocal of the second fraction. The *reciprocal* of the fraction $\frac{x}{y}$ is the fraction $\frac{y}{x}$. Here is an example:

$$\frac{1}{3} \div \frac{5}{6} = \frac{1}{3} \times \frac{6}{5} = \frac{6}{15} = \frac{2}{5}$$

Factorization

Factors are the numbers multiplied to achieve a product. Thus, every product in a multiplication equation has, at minimum, two factors. Of course, some products will have more than two factors. For the sake of most discussions, assume that factors are positive integers.

To find a number's factors, start with 1 and the number itself. Then divide the number by 2, 3, 4, and so on, seeing if any divisors can divide the number without a remainder, keeping a list of those that do. Stop upon reaching either the number itself or another factor.

Let's find the factors of 45. Start with 1 and 45. Then try to divide 45 by 2, which fails. Now divide 45 by 3. The answer is 15, so 3 and 15 are now factors. Dividing by 4 doesn't work, and dividing by 5 leaves 9. Lastly, dividing 45 by 6, 7, and 8 all don't work. The next integer to try is 9, but this is already known to be a factor, so the factorization is complete. The factors of 45 are 1, 3, 5, 9, 15 and 45.

<u>Prime Factorization</u>

Prime factorization involves an additional step after breaking a number down to its factors: breaking down the factors until they are all prime numbers. A prime number is any number that can only be divided by 1 and itself. The prime numbers between 1 and 20 are 2, 3, 5, 7, 11, 13, 17, and 19. As a simple test, numbers that are even or end in 5 are not prime.

Let's break 129 down into its prime factors. First, the factors are 3 and 43. Both 3 and 43 are prime numbers, so we're done. But if 43 was not a prime number, then it would also need to be factorized until all of the factors are expressed as prime numbers.

Common Factor

A common factor is a factor shared by two numbers. Let's take 45 and 30 and find the common factors:

The factors of 45 are: 1, 3, 5, 9, 15, and 45.
The factors of 30 are: 1, 2, 3, 5, 6, 10, 15, and 30.
The common factors are 1, 3, 5, and 15.

Greatest Common Factor

The greatest common factor is the largest number among the shared, common factors. From the factors of 45 and 30, the common factors are 3, 5, and 15. Thus, 15 is the greatest common factor, as it's the largest number.

Least Common Multiple

The least common multiple is the smallest number that's a multiple of two numbers. Let's try to find the least common multiple of 4 and 9. The multiples of 4 are 4, 8, 12, 16, 20, 24, 28, 32, 36, and so on. For 9, the multiples are 9, 18, 27, 36, 45, 54, etc. Thus, the least common multiple of 4 and 9 is 36, the lowest number where 4 and 9 share multiples.

If two numbers share no factors besides 1 in common, then their least common multiple will be simply their product. If two numbers have common factors, then their least common multiple will be their product divided by their greatest common factor. This can be visualized by the formula $LCM = \frac{x \times y}{GCF}$, where *x* and *y* are some integers and *LCM* and *GCF* are their least common multiple and greatest common factor, respectively.

Fractions

A fraction is an equation that represents a part of a whole, but can also be used to present ratios or division problems. An example of a fraction is $\frac{x}{y}$. In this example, x is called the numerator, while y is the denominator. The numerator represents the number of parts, and the denominator is the total number of parts. They are separated by a line or slash. In simple fractions, the numerator and denominator can be nearly any integer. However, the denominator of a fraction can never be zero, because dividing by zero is a function which is undefined.

Imagine that an apple pie has been baked for a holiday party, and the full pie has eight slices. After the party, there are five slices left. How could the amount of the pie that remains be expressed as a fraction? The numerator is 5 since there are 5 pieces left, and the denominator is 8 since there were eight total slices in the whole pie. Thus, expressed as a fraction, the leftover pie totals $\frac{5}{8}$ of the original amount.

Fractions come in three different varieties: proper fractions, improper fractions, and mixed numbers. Proper fractions have a numerator less than the denominator, such as $\frac{3}{8}$, but improper fractions have a numerator greater than the denominator, such as $\frac{15}{8}$. Mixed numbers combine a whole number with a proper fraction, such as $3\frac{1}{2}$. Any mixed number can be written as an improper fraction by multiplying the integer by the denominator, adding the product to the value of the numerator, and dividing the sum by the original denominator. For example, $3\frac{1}{2} = \frac{3 \times 2 + 1}{2} = \frac{7}{2}$. Whole numbers can also be converted into fractions by placing the whole number as the numerator and making the denominator 1. For example, $3 = \frac{3}{1}$.

One of the most fundamental concepts of fractions is their ability to be manipulated by multiplication or division. This is possible since $\frac{n}{n} = 1$ for any non-zero integer. As a result, multiplying or dividing by $\frac{n}{n}$ will not alter the original fraction since any number multiplied or divided by 1 doesn't change the value of that number. Fractions of the same value are known as equivalent fractions. For example, $\frac{2}{4}, \frac{4}{8}, \frac{50}{100}, \text{and} \frac{75}{150}$ are equivalent, as they all equal $\frac{1}{2}$.

Although many equivalent fractions exist, they are easier to compare and interpret when reduced or simplified. The numerator and denominator of a simple fraction will have no factors in common other than 1. When reducing or simplifying fractions, divide the numerator and denominator by the greatest common factor. A simple strategy is to divide the numerator and denominator by low numbers, like 2, 3, or 5 until arriving at a simple fraction, but the same thing could be achieved by determining the greatest common factor for both the numerator and denominator and dividing each by it. Using the first method is preferable when both the numerator and denominator are even, end in 5, or are obviously a multiple of another number. However, if no numbers seem to work, it will be necessary to factor the numerator and denominator to find the GCF. Let's look at examples:

1) Simplify the fraction $\frac{6}{8}$:

Dividing the numerator and denominator by 2 results in $\frac{3}{4}$, which is a simple fraction.

2) Simplify the fraction $\frac{12}{36}$:

Dividing the numerator and denominator by 2 leaves $\frac{6}{18}$. This isn't a simple fraction, as both the numerator and denominator have factors in common. Diving each by 3 results in $\frac{2}{6}$, but this can be further simplified by dividing by 2 to get $\frac{1}{3}$. This is the simplest fraction, as the numerator is 1. In cases like this, multiple division operations can be avoided by determining the greatest common factor between the numerator and denominator.

3) Simplify the fraction $\frac{18}{54}$ by dividing by the greatest common factor:

First, determine the factors for the numerator and denominator. The factors of 18 are 1, 2, 3, 6, 9, and 18. The factors of 54 are 1, 2, 3, 6, 9, 18, 27, and 54. Thus, the greatest common factor is 18. Dividing $\frac{18}{54}$ by 18 leaves $\frac{1}{3}$, which is the simplest fraction. This method takes slightly more work, but it definitively arrives at the simplest fraction.

Operations with Fractions

Of the four basic operations that can be performed on fractions, the one which involves the least amount of work is multiplication. To multiply two fractions, simply multiply the numerators, multiply the denominators, and place the products as a fraction. Whole numbers and mixed numbers can also be expressed as a fraction, as described above, to multiply with a fraction. Let's work through a couple of examples.

1) $\frac{2}{5} \times \frac{3}{4} = \frac{6}{20} = \frac{3}{10}$

2) $\frac{4}{9} \times \frac{7}{11} = \frac{28}{99}$

Dividing fractions is similar to multiplication with one key difference. To divide fractions, flip the numerator and denominator of the second fraction, and then proceed as if it were a multiplication problem:

1) $\frac{7}{8} \div \frac{4}{5} = \frac{7}{8} \times \frac{5}{4} = \frac{35}{32}$

2) $\frac{5}{9} \div \frac{1}{3} = \frac{5}{9} \times \frac{3}{1} = \frac{15}{9} = \frac{5}{3}$

Addition and subtraction require more steps than multiplication and division, as these operations require the fractions to have the same denominator, also called a common denominator. It is always possible to find a common denominator by multiplying the denominators. However, when the denominators are large numbers, this method is unwieldy, especially if the answer must be provided in its simplest form. Thus, it's beneficial to find the least common denominator of the fractions—the least common denominator is incidentally also the least common multiple.

Once equivalent fractions have been found with common denominators, simply add or subtract the numerators to arrive at the answer:

1) $\frac{1}{2} + \frac{3}{4} = \frac{2}{4} + \frac{3}{4} = \frac{5}{4}$

2) $\frac{3}{12} + \frac{11}{20} = \frac{15}{60} + \frac{33}{60} = \frac{48}{60} = \frac{4}{5}$

3) $\frac{7}{9} - \frac{4}{15} = \frac{35}{45} - \frac{12}{45} = \frac{23}{45}$

4) $\frac{5}{6} - \frac{7}{18} = \frac{15}{18} - \frac{7}{18} = \frac{8}{18} = \frac{4}{9}$

Recognizing Equivalent Fractions and Mixed Numbers

The value of a fraction does not change if multiplying or dividing both the numerator and the denominator by the same number (other than 0). In other words, $\frac{x}{y} = \frac{a \times x}{a \times y} = \frac{x \div a}{y \div a}$, as long as *a* is not 0. This means that $\frac{2}{5} = \frac{4}{10}$, for example. If *x* and *y* are integers that have no common factors, then the fraction is said to be *simplified*. This means $\frac{2}{5}$ is simplified, but $\frac{4}{10}$ is not.

Often when working with fractions, the fractions need to be rewritten so that they all share a single denominator—this is called finding a *common denominator* for the fractions. Using two fractions, $\frac{a}{b}$ and $\frac{c}{d}$, the numerator and denominator of the left fraction can be multiplied by *d*, while the numerator and denominator of the right fraction can be multiplied by *b*. This provides the fractions $\frac{a \times d}{b \times d}$ and $\frac{c \times b}{d \times b}$ with the common denominator $b \times d$.

A fraction whose numerator is smaller than its denominator is called a *proper fraction*. A fraction whose numerator is bigger than its denominator is called an *improper fraction*. These numbers can be rewritten as a combination of integers and fractions, called a *mixed number*. For example, $\frac{6}{5} = \frac{5}{5} + \frac{1}{5} = 1 + \frac{1}{5}$, and can be written as $1\frac{1}{5}$.

Applying Estimation Strategies and Rounding Rules to Real-World Problems

Estimation

Estimation is finding a value that is close to a solution but is not the exact answer. For example, if there are values in the thousands to be multiplied, then each value can be estimated to the nearest thousand and the calculation performed. This value provides an approximate solution that can be determined very quickly.

Rounding Numbers

It's often convenient to round a number, which means to give an approximate figure to make it easier to compare amounts or perform mental math. Round up when the digit is 5 or more. The rounded digit, and all subsequent digits, becomes 0, and the preceding digit goes up by 1. Here are some examples:

75 rounded to the nearest ten is 80
380 rounded to the nearest hundred is 400
22.697 rounded to the nearest hundredth is 22.70

Round down when rounding on any digit that is below 5. The rounded digit, and all subsequent digits, becomes 0, and the preceding digit goes down by 1. Here are some examples:

92 rounded to the nearest ten is 90
839 rounded to the nearest hundred is 800
22.643 rounded to the nearest hundredth is 22.64

Determining the Reasonableness of Results

When solving math word problems, the solution obtained should make sense within the given scenario. The step of checking the solution will reduce the possibility of a calculation error or a solution that may be *mathematically* correct but not applicable in the real world. Consider the following scenarios:

A problem states that Lisa got 24 out of 32 questions correct on a test and asks to find the percentage of correct answers. To solve the problem, a student divided 32 by 24 to get 1.33, and then multiplied by 100 to get 133 percent. By examining the solution within the context of the problem, the student should recognize that getting all 32 questions correct will produce a perfect score of 100 percent. Therefore, a score of 133 percent with 8 incorrect answers does not make sense and the calculations should be checked.

A problem states that the maximum weight on a bridge cannot exceed 22,000 pounds. The problem asks to find the maximum number of cars that can be on the bridge at one time if each car weighs 4,000 pounds. To solve this problem, a student divided 22,000 by 4,000 to get an answer of 5.5. By examining the solution within the context of the problem, the student should recognize that although the calculations are mathematically correct, the solution does not make sense. Half of a car on a bridge is not possible, so the student should determine that a maximum of 5 cars can be on the bridge at the same time.

Mental Math Estimation

Once a result is determined to be logical within the context of a given problem, the result should be evaluated by its nearness to the expected answer. This is performed by approximating given values to perform mental math. Numbers should be rounded to the nearest value possible to check the initial results.

Consider the following example: A problem states that a customer is buying a new sound system for their home. The customer purchases a stereo for $435, 2 speakers for $67 each, and the necessary cables for $12. The customer chooses an option that allows him to spread the costs over equal payments for 4 months. How much will the monthly payments be?

After making calculations for the problem, a student determines that the monthly payment will be $145.25. To check the accuracy of the results, the student rounds each cost to the nearest ten ($440 + 70 + 70 + 10$) and determines that the total is approximately $590. Dividing by 4 months gives an approximate monthly payment of $147.50. Therefore, the student can conclude that the solution of $145.25 is very close to what should be expected.

When rounding, the place-value that is used in rounding can make a difference. Suppose the student had rounded to the nearest hundred for the estimation. The result ($400 + 100 + 100 + 0 = 600;\ 600 \div 4 = 150$) will show that the answer is reasonable, but not as close to the actual value as rounding to the nearest ten.

Operations with Decimals and Percents

Recognition of Decimals

The *decimal system* is a way of writing out numbers that uses ten different numerals: 0, 1, 2, 3, 4, 5, 6, 7, 8, and 9. This is also called a "base ten" or "base 10" system. Other bases are also used. For example, computers work with a base of 2. This means they only use the numerals 0 and 1.

The *decimal place* denotes how far to the right of the decimal point a numeral is. The first digit to the right of the decimal point is in the *tenths* place. The next is the *hundredths*. The third is the *thousandths*.

So, 3.142 has a 1 in the tenths place, a 4 in the hundredths place, and a 2 in the thousandths place.

The *decimal point* is a period used to separate the *ones* place from the *tenths* place when writing out a number as a decimal.

A *decimal number* is a number written out with a decimal point instead of as a fraction, for example, 1.25 instead of $\frac{5}{4}$. Depending on the situation, it can sometimes be easier to work with fractions and sometimes easier to work with decimal numbers.

A decimal number is *terminating* if it stops at some point. It is called *repeating* if it never stops, but repeats a pattern over and over. It is important to note that every rational number can be written as a terminating decimal or as a repeating decimal.

Addition with Decimals

To add decimal numbers, each number in columns needs to be lined up by the decimal point. For each number being added, the zeros to the right of the last number need to be filled in so that each of the numbers has the same number of places to the right of the decimal. Then, the columns can be added together. Here is an example of 2.45 + 1.3 + 8.891 written in column form:

2.450

1.300

+ 8.891

Zeros have been added in the columns so that each number has the same number of places to the right of the decimal.

Added together, the correct answer is 12.641:

2.450

1.300

+ 8.891

12.641

Subtraction with Decimals

Subtracting decimal numbers is the same process as adding decimals. Here is 7.89 – 4.235 written in column form:

7.890

- 4.235

3.655

A zero has been added in the column so that each number has the same number of places to the right of the decimal.

Multiplication with Decimals

Decimals can be multiplied as if there were no decimals points in the problem. For example, 0.5 x 1.25 can be rewritten and multiplied as 5 x 125, which equals 625.

The final answer will have the same number of decimal *points* as the total number of decimal *places* in the problem. The first number has one decimal place, and the second number has two decimal places. Therefore, the final answer will contain three decimal places:

0.5 x 1.25 = 0.625

Division with Decimals

Dividing a decimal by a whole number entails using long division first by ignoring the decimal point. Then, the decimal point is moved the number of places given in the problem.

For example, $6.8 \div 4$ can be rewritten as $68 \div 4$, which is 17. There is one non-zero integer to the right of the decimal point, so the final solution would have one decimal place to the right of the solution. In this case, the solution is 1.7.

Dividing a decimal by another decimal requires changing the divisor to a whole number by moving its decimal point. The decimal place of the dividend should be moved by the same number of places as the divisor. Then, the problem is the same as dividing a decimal by a whole number.

For example, $5.72 \div 1.1$ has a divisor with one decimal point in the denominator. The expression can be rewritten as $57.2 \div 11$ by moving each number one decimal place to the right to eliminate the decimal. The long division can be completed as $572 \div 11$ with a result of 52. Since there is one non-zero integer to the right of the decimal point in the problem, the final solution is 5.2.

In another example, $8 \div 0.16$ has a divisor with two decimal points in the denominator. The expression can be rewritten as $800 \div 16$ by moving each number two decimal places to the right to eliminate the decimal in the divisor. The long division can be completed with a result of 50.

Percentages

Think of percentages as fractions with a denominator of 100. In fact, percentage means "per hundred." Problems often require converting numbers from percentages, fractions, and decimals.

Percent Problems

The basic percent equation is the following:

$$\frac{is}{of} = \frac{\%}{100}$$

The placement of numbers in the equation depends on what the question asks.

Example 1

Find 40% of 80.

Basically, the problem is asking, "What is 40% of 80?" The 40% is the percent, and 80 is the number to find the percent "of." The equation is:

$$\frac{x}{80} = \frac{40}{100}$$

Solving the equation by cross-multiplication, the problem becomes 100x = 80(40). Solving for x gives the answer: x = 32.

Example 2

What percent of 100 is 20?

The 20 fills in the "is" portion, while 100 fills in the "of." The question asks for the percent, so that will be x, the unknown. The following equation is set up:

$$\frac{20}{100} = \frac{x}{100}$$

Cross-multiplying yields the equation 100x = 20(100). Solving for x gives the answer of 20%.

Example 3
30% of what number is 30?

The following equation uses the clues and numbers in the problem:

$$\frac{30}{x} = \frac{30}{100}$$

Cross-multiplying results in the equation 30(100) = 30x. Solving for x gives the answer x = 100.

Conversions

Decimals and Percentages
Since a percentage is based on "per hundred," decimals and percentages can be converted by multiplying or dividing by 100. Practically speaking, this always amounts to moving the decimal point two places to the right or left, depending on the conversion. To convert a percentage to a decimal, move the decimal point two places to the left and remove the % sign. To convert a decimal to a percentage, move the decimal point two places to the right and add a "%" sign. Here are some examples:

65% = 0.65
0.33 = 33%
0.215 = 21.5%
99.99% = 0.9999
500% = 5.00
7.55 = 755%

Fractions and Percentages
Remember that a percentage is a number per one hundred. So a percentage can be converted to a fraction by making the number in the percentage the numerator and putting 100 as the denominator:

$$43\% = \frac{43}{100}$$

$$97\% = \frac{97}{100}$$

Note that the percent symbol (%) kind of looks like a 0, a 1, and another 0. So think of a percentage like 54% as 54 over 100.

To convert a fraction to a percent, follow the same logic. If the fraction happens to have 100 in the denominator, you're in luck. Just take the numerator and add a percent symbol:

$$\frac{28}{100} = 28\%$$

Otherwise, divide the numerator by the denominator to get a decimal:

$$\frac{9}{12} = 0.75$$

Then convert the decimal to a percentage:

$$0.75 = 75\%$$

Another option is to make the denominator equal to 100. Be sure to multiply the numerator by the same number as the denominator. For example:

$$\frac{3}{20} \times \frac{5}{5} = \frac{15}{100}$$

$$\frac{15}{100} = 15\%$$

Changing Fractions to Decimals

To change a fraction into a decimal, divide the denominator into the numerator until there are no remainders. There may be repeating decimals, so rounding is often acceptable. A straight line above the repeating portion denotes that the decimal repeats.

Example

Express 4/5 as a decimal.

Set up the division problem.

$$5\overline{)4}$$

5 does not go into 4, so place the decimal and add a zero.

$$5\overline{)4.0}$$

5 goes into 40 eight times. There is no remainder.

$$\begin{array}{r} 0.8 \\ 5\overline{)4.0} \\ -4.0 \\ \hline 0 \end{array}$$

The solution is 0.8.

Example

Express 33 1/3 as a decimal.

Since the whole portion of the number is known, set it aside to calculate the decimal from the fraction portion.

Set up the division problem.

$$3\overline{)1}$$

3 does not go into 1, so place the decimal and add zeros. 3 goes into 10 three times.

$$3\overline{)1.000}$$

This will repeat with a remainder of 3, so place a line over the 3 denotes the repetition.

$$
\begin{array}{r}
0.333 \\
3\overline{)1.000} \\
-9 \\
\hline
10 \\
-9 \\
\hline
10
\end{array}
$$

The solution is $0.\bar{3}$

<u>Changing Decimals to Fractions</u>
To change decimals to fractions, place the decimal portion of the number, the numerator, over the respective place value, the denominator, then reduce, if possible.

Example
Express 0.25 as a fraction.

This is read as twenty-five hundredths, so put 25 over 100. Then reduce to find the solution.

$$\frac{25}{100} = \frac{1}{4}$$

Example
Express 0.455 as a fraction

This is read as four hundred fifty-five thousandths, so put 455 over 1000. Then reduce to find the solution.

$$\frac{455}{1000} = \frac{91}{200}$$

There are two types of problems that commonly involve percentages. The first is to calculate some percentage of a given quantity, where you convert the percentage to a decimal, and multiply the quantity by that decimal. Secondly, you are given a quantity and told it is a fixed percent of an unknown quantity. In this case, convert to a decimal, then divide the given quantity by that decimal.

Example
What is 30% of 760?

Convert the percent into a useable number. "Of" means to multiply.

$$30\% = 0.30$$

Set up the problem based on the givens, and solve.

$$0.30 \times 760 = 228$$

Example
8.4 is 20% of what number?

Convert the percent into a useable number.

$$20\% = 0.20$$

The given number is a percent of the answer needed, so divide the given number by this decimal rather than multiplying it.

$$\frac{8.4}{0.20} = 42$$

Applications and Problem-Solving

Solving for X in Proportions

Proportions are commonly used to solve word problems to find unknown values such as *x* that are some percent or fraction of a known number. Proportions are solved by cross-multiplying and then dividing to arrive at x. The following examples show how this is done:

1) $\frac{75\%}{90\%} = \frac{25\%}{x}$

To solve for x, the fractions must be cross multiplied: $(75\%x = 90\% \times 25\%)$. To make things easier, let's convert the percentages to decimals: $(0.9 \times 0.25 = 0.225 = 0.75x)$. To get rid of x's co-efficient, each side must be divided by that same coefficient to get the answer $x = 0.3$. The question could ask for the answer as a percentage or fraction in lowest terms, which are 30% and $\frac{3}{10}$, respectively.

2) $\frac{x}{12} = \frac{30}{96}$

Cross-multiply: $96x = 30 \times 12$
Multiply: $96x = 360$
Divide: $x = 360 \div 96$
Answer: $x = 3.75$

3) $\frac{0.5}{3} = \frac{x}{6}$

Cross-multiply: $3x = 0.5 \times 6$
Multiply: $3x = 3$
Divide: $x = 3 \div 3$
Answer: $x = 1$

You may have noticed there's a faster way to arrive at the answer. If there is an obvious operation being performed on the proportion, the same operation can be used on the other side of the proportion to solve for x. For example, in the first practice problem, 75% became 25% when divided by 3, and upon doing the same to 90%, the correct answer of 30% would have been found with much less legwork. However, these questions aren't always so intuitive, so it's a good idea to work through the steps, even if the answer seems apparent from the outset.

There is a specific order in which operations are performed:

Parentheses – calculate anything inside parentheses first.

Exponents – apply any exponents second.

Multiplication – execute any multiplication.

Division – execute any division.

Addition – execute any addition.

Subtraction – execute any subtraction.

A memory device to help recall the order is word PEMDAS, or "Please Excuse My Dear Aunt Sally."

Example
Solve using correct order of operations $(3 + 4)(4 \div 2) + 8$.

Calculate anything inside parentheses first.

$$(3 + 4)(4 \div 2) + 8$$

Multiply.

$$(7)(2) + 8$$

Add, then solve.

$$14 + 8 = 22$$

Not every equation contains every operator, but the order of operations needs to be followed to obtain the correct answer. Some tell you the number that represents a variable. In that case, replace the variable with the number first, then follow the order to solve.

Example
Solve $X^2 + 5 - 1$, for $X = 3$.

Replace X with 3.

$$X^2 + 5 - 1$$

Use order of operations to solve exponents.

$$3^2 + 5 - 1$$

Add and subtract.

$$9 + 5 - 1$$

Solve.

$$14 - 1 = 13$$

Another application for algebra is temperature conversion. The United States commonly uses the Fahrenheit scale to measure temperature. In science and medicine, the Celsius scale is used. Here are some common comparisons:

$0\,^0C = 32\,^0F$, water freezes.

$100\,^0C = 212\,^0F$, water boils.

To convert between the two temperature scales, use the following equations.

$$9/5\,^0C + 32 = {}^0F$$

$$5/9\,(^0F - 32) = {}^0C$$

Example

A patient has a temperature of 39.6 ^{0}C. Convert this to Fahrenheit to assess whether she needs medical care.

Set up the equation using 39.6 for ^{0}C.

$$9/5\,(39.6) + 32 = {}^0F$$

Multiply, divide and then add.

$$71.28 + 32 = {}^0F$$

Solve.

$$103.28\,^0F$$

With a temperature over 103 ^{0}F, the patient may need medical care.

When trying to isolate a term or solve for a variable on one side of an equation, it is important not to change the equation. Always do the same operations to both sides of the equation.

Example

Solve for X, $X - 9 = 10$.

Solve for X by isolating it on a side.

$$X - 9 = 10$$

To get X alone, eliminate the 9 by adding 9 to both sides.

$$X - 9 + 9 = 10 + 9$$

Solve.

$$X = 19$$

Example
Solve for X, $4X = 20$.

Solve for X by isolating it on one side.

$$4X = 20$$

To get X alone, eliminate the 4 by dividing both sides by 4.

$$\frac{4x}{4} = \frac{20}{4}$$

Solve

$$X = 5$$

Example
Solve for X, $X^2 - 2 = 7$.

Isolate X on a side by adding 2 to both sides.

$$X^2 - 2 + 2 = 7 + 2$$

To undo the squaring of X, take the square root of both sides.

$$\sqrt{X^2} = \sqrt{9}$$

Solve.

$$X = 3$$

While solving an equation, you can also combine like terms. This is also called simplifying.

Examples of like terms would be X^2 and $3X^2$, or 4X and 8X.

Simplify: $X^2 + 2X^2 + 9X - 3X + 1 - 5$. This is not a full equation so we cannot solve it, only simplify it.

Identify all like terms.

$$X^2 + 2X^2 + 9X - 3X + 1 - 5$$

Combine the terms. Be sure to use the proper signs.

$$3X^2 + 6X - 4$$

Ratio Problems
A *ratio* compares the size of one group to the size of another. For example, there may be a room with 4 tables and 24 chairs. The ratio of tables to chairs is 4: 24. Such ratios behave like fractions in that both sides of the ratio by the same number can be multiplied or divided. Thus, the ratio 4:24 is the same as the ratio 2:12 and 1:6.

One quantity is *proportional* to another quantity if the first quantity is always some multiple of the second. For instance, the distance travelled in five hours is always five times to the speed as travelled. The distance is proportional to speed in this case.

One quantity is *inversely proportional* to another quantity if the first quantity is equal to some number divided by the second quantity. The time it takes to travel one hundred miles will be given by 100 divided by the speed travelled. The time is inversely proportional to the speed.

When dealing with word problems, there is no fixed series of steps to follow, but there are some general guidelines to use. It is important that the quantity that to be found is identified. Then, it can be determined how the given values can be used and manipulated to find the final answer.

Example 1
Jana wants to travel to visit Alice, who lives one hundred and fifty miles away. If she can drive at fifty miles per hour, how long will her trip take?

The quantity to find is the *time* of the trip. The time of a trip is given by the distance to travel divided by the speed to be traveled. The problem determines that the distance is one hundred and fifty miles, while the speed is fifty miles per hour. Thus, 150 divided by 50 is $150 \div 50 = 3$. Because *miles* and *miles per hour* are the units being divided, the miles cancel out. The result is 3 hours.

Example 2
Bernard wishes to paint a wall that measures twenty feet wide by eight feet high. It costs ten cents to paint one square foot. How much money will Bernard need for paint?

The final quantity to compute is the *cost* to paint the wall. This will be ten cents ($0.10) for each square foot of area needed to paint. The area to be painted is unknown, but the dimensions of the wall are given; thus, it can be calculated.

The dimensions of the wall are 20 feet wide and 8 feet high. Since the area of a rectangle is length multiplied by width, the area of the wall is 8 x 20 = 160 square feet. Multiplying 0.1 x 160 yields $16 as the cost of the paint.

The *average* or *mean* of a collection of numbers is given by adding those numbers together and then dividing by the total number of values. A *weighted average* or *weighted mean* is given by adding the numbers multiplied by their weights, then dividing by the sum of the weights:

$$\frac{w_1 x_1 + w_2 x_2 + w_3 x_3 \ldots + w_n x_n}{w_1 + w_2 + w_3 + \cdots + w_n}$$

An *ordinary average* is a weighted average where all the weights are 1.

Fraction and Percent Equivalencies

The word *percent* comes from the Latin phrase for "per one hundred." A *percent* is a way of writing out a fraction. It is a fraction with a denominator of 100. Thus, $65\% = \frac{65}{100}$.

To convert a fraction to a percent, the denominator is written as 100. For example, $\frac{3}{5} = \frac{60}{100} = 60\%$.

In converting a percent to a fraction, the percent is written with a denominator of 100, and the result is simplified. For example, $30\% = \frac{30}{100} = \frac{3}{10}$.

Distribution of a Quantity into its Fractional Parts

A quantity may be broken into its fractional parts. For example, a toy box holds three types of toys for kids. $\frac{1}{3}$ of the toys are Type A and $\frac{1}{4}$ of the toys are Type B. With that information, how many Type C toys are there?

First, the sum of Type A and Type B must be determined by finding a common denominator to add the fractions. The lowest common multiple is 12, so that is what will be used. The sum is $\frac{1}{3}+\frac{1}{4}=\frac{4}{12}+\frac{3}{12}=\frac{7}{12}$.

This value is subtracted from 1 to find the number of Type C toys. The value is subtracted from 1 because 1 represents a whole. The calculation is $1-\frac{7}{12}=\frac{12}{12}-\frac{7}{12}=\frac{5}{12}$. This means that $\frac{5}{12}$ of the toys are Type C. To check the answer, add all fractions together, and the result should be 1.

Simple Geometry Problems

There are many key facts related to geometry that are applicable. The sum of the measures of the angles of a triangle are 180°, and for a quadrilateral, the sum is 360°. Rectangles and squares each have four right angles. A *right angle* has a measure of 90°.

Perimeter

The *perimeter* is the distance around a figure or the sum of all sides of a polygon.

The *formula for the perimeter of a square* is four times the length of a side. For example, the following square has side lengths of 5 meters:

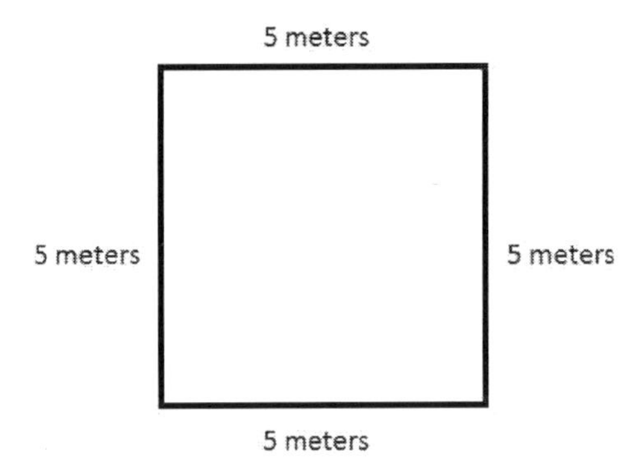

The perimeter is 20 meters because 4 times 5 is 20.

The *formula for a perimeter of a rectangle* is the sum of twice the length and twice the width. For example, if the length of a rectangle is 10 inches and the width 8 inches, then the perimeter is 36 inches because $P=2l+2w=2(10)+2(8)=20+16=36$ inches.

Area

The area is the amount of space inside of a figure, and there are formulas associated with area.

The area of a triangle is the product of one-half the base and height. For example, if the base of the triangle is 2 feet and the height 4 feet, then the area is 4 square feet. The following equation shows the formula used to calculate the area of the triangle:

$$A = \frac{1}{2}bh = \frac{1}{2}(2)(4) = 4 \text{ square feet}$$

The area of a square is the length of a side squared, and the area of a rectangle is length multiplied by the width. For example, if the length of the square is 7 centimeters, then the area is 49 square centimeters. The formula for this example is $A = s^2 = 7^2 = 49$ square centimeters. An example is if the rectangle has a length of 6 inches and a width of 7 inches, then the area is 42 square inches:

$$A = lw = 6(7) = 42 \text{ square inches}$$

The area of a trapezoid is one-half the height times the sum of the bases. For example, if the length of the bases are 2.5 and 3 feet and the height 3.5 feet, then the area is 9.625 square feet. The following formula shows how the area is calculated:

$$A = \frac{1}{2}h(b_1 + b_2) = \frac{1}{2}(3.5)(2.5 + 3) = \frac{1}{2}(3.5)(5.5) = 9.625 \text{ square feet}$$

The perimeter of a figure is measured in single units, while the area is measured in square units.

Practice Questions

1. 3.4+2.35+4=
   a. 5.35
   b. 9.2
   c. 9.75
   d. 10.25

2. $5.88 \times 3.2 =$
   a. 18.816
   b. 16.44
   c. 20.352
   d. 17

3. $\frac{3}{25} =$
   a. 0.15
   b. 0.1
   c. 0.9
   d. 0.12

4. Which of the following is largest?
   a. 0.45
   b. 0.096
   c. 0.3
   d. 0.313

5. Which of the following is NOT a way to write 40 percent of *N*?
   a. $(0.4)N$
   b. $\frac{2}{5}N$
   c. $40N$
   d. $\frac{4N}{10}$

6. Which is closest to 17.8×9.9?
   a. 140
   b. 180
   c. 200
   d. 350

7. A student gets an 85% on a test with 20 questions. How many answers did the student solve correctly?
   a. 15
   b. 16
   c. 17
   d. 18

8. Four people split a bill. The first person pays for $\frac{1}{5}$, the second person pays for $\frac{1}{4}$, and the third person pays for $\frac{1}{3}$. What fraction of the bill does the fourth person pay?

a. $\frac{13}{60}$
b. $\frac{47}{60}$
c. $\frac{1}{4}$
d. $\frac{4}{15}$

9. 6 is 30% of what number?

a. 18
b. 20
c. 24
d. 26

10. $3\frac{2}{3} - 1\frac{4}{5} =$

a. $1\frac{13}{15}$
b. $\frac{14}{15}$
c. $2\frac{2}{3}$
d. $\frac{4}{5}$

11. What is $\frac{420}{98}$ rounded to the nearest integer?

a. 4
b. 3
c. 5
d. 6

12. $4\frac{1}{3} + 3\frac{3}{4} =$

a. $6\frac{5}{12}$
b. $8\frac{1}{12}$
c. $8\frac{2}{3}$
d. $7\frac{7}{12}$

13. Five of six numbers have a sum of 25. The average of all six numbers is 6. What is the sixth number?

a. 8
b. 10
c. 11
d. 12

14. $52.3 \times 10^{-3} =$

a. 0.00523
b. 0.0523
c. 0.523
d. 523

15. If $\frac{5}{2} \div \frac{1}{3} = n$, then n is between:

a. 5 and 7
b. 7 and 9
c. 9 and 11
d. 3 and 5

16. A closet is filled with red, blue, and green shirts. If $\frac{1}{3}$ of the shirts are green and $\frac{2}{5}$ are red, what fraction of the shirts are blue?

a. $\frac{4}{15}$
b. $\frac{1}{5}$
c. $\frac{7}{15}$
d. $\frac{1}{2}$

17. Shawna buys $2\frac{1}{2}$ gallons of paint. If she uses $\frac{1}{3}$ of it on the first day, how much does she have left?

a. $1\frac{5}{6}$ gallons

b. $1\frac{1}{2}$ gallons

c. $1\frac{2}{3}$gallons

d. 2 gallons

18. How will $\frac{4}{5}$ be written as a percent?

a. 40%
b. 125%
c. 90%
d. 80%

19. What are all the factors of 12?

a. 12, 24, 36
b. 1, 2, 4, 6, 12
c. 12, 24, 36, 48
d. 1, 2, 3, 4, 6, 12

20. At the beginning of the day, Xavier has 20 apples. At lunch, he meets his sister Emma and gives her half of his apples. After lunch, he stops by his neighbor Jim's house and gives him 6 of his apples. He then uses ¾ of his remaining apples to make an apple pie for dessert at dinner. At the end of the day, how many apples does Xavier have left?

a. 4
b. 6
c. 2
d. 1

21. How will the number 847.89632 be written if rounded to the nearest hundredth?
   a. 847.90
   b. 900
   c. 847.89
   d. 847.896

22. What is the value of the sum of $\frac{1}{3}$ and $\frac{2}{5}$?
   a. $\frac{3}{8}$
   b. $\frac{11}{15}$
   c. $\frac{11}{30}$
   d. $\frac{4}{5}$

23. Add and express in reduced form $5/12 + 4/9$.
   a. 9/17
   b. 1/3
   c. 31/36
   d. 3/5

24. Divide and reduce $4/13 \div 27/169$.
   a. 52/27
   b. 51/27
   c. 52/29
   d. 51/29

25. Express as a reduced mixed number $54/15$.
   a. 3 3/5
   b. 3 1/15
   c. 3 3/54
   d. 3 1/54

26. In the problem $5 \times 6 + 4 \div 2 - 1$, which operation should be completed first?
   a. Multiplication
   b. Addition
   c. Division
   d. Subtraction

27. Express as an improper fraction $8\ 3/7$.
   a. 11/7
   b. 21/8
   c. 5/3
   d. 59/7

28. Express as an improper fraction 11 5/8.

a. 55/8

b. 93/8

c. 16/11

d. 19/5

29. Round to the nearest tenth 8.067.

a. 8.07

b. 8.10

c. 8.00

d. 8.11

30. When rounding 245.2678 to the nearest thousandth, which place value would be used to decide whether to round up or round down?

a. Ten-thousandth

b. Thousandth

c. Hundredth

d. Thousand

Answer Explanations

1. C: The decimal points are lined up, with zeroes put in as needed. Then, the numbers are added just like integers:

$$\begin{array}{r} 3.40 \\ 2.35 \\ +\underline{4.00} \\ 9.75 \end{array}$$

2. A: This problem can be multiplied as 588×32, except at the end, the decimal point needs to be moved three places to the left. Performing the multiplication will give 18,816, and moving the decimal place over three places results in 18.816.

3. D: The fraction is converted so that the denominator is 100 by multiplying the numerator and denominator by 4, to get $\frac{3}{25} = \frac{12}{100}$. Dividing a number by 100 just moves the decimal point two places to the left, with a result of 0.12.

4. A: Figure out which is largest by looking at the first non-zero digits. Choice *B*'s first non-zero digit is in the hundredths place. The other three all have non-zero digits in the tenths place, so it must be *A*, *C*, or *D*. Of these, *A* has the largest first non-zero digit.

5. C: 40*N* would be 4000% of *N*. It's possible to check that each of the others is actually 40% of *N*.

6. B: Instead of multiplying these out, the product can be estimated by using $18 \times 10 = 180$. The error here should be lower than 15, since it is rounded to the nearest integer, and the numbers add to something less than 30.

7. C: 85% of a number means multiplying that number by 0.85. So, $0.85 \times 20 = \frac{85}{100} \times \frac{20}{1}$, which can be simplified to $\frac{17}{20} \times \frac{20}{1} = 17$.

8. A: To find the fraction of the bill that the first three people pay, the fractions need to be added, which means finding common denominator. The common denominator will be 60. $\frac{1}{5} + \frac{1}{4} + \frac{1}{3} = \frac{12}{60} + \frac{15}{60} + \frac{20}{60} = \frac{47}{60}$. The remainder of the bill is $1 - \frac{47}{60} = \frac{60}{60} - \frac{47}{60} = \frac{13}{60}$.

9. B: 30% is 3/10. The number itself must be 10/3 of 6, or $\frac{10}{3} \times 6 = 10 \times 2 = 20$.

10. A: These numbers to improper fractions: $\frac{11}{3} - \frac{9}{5}$. Take 15 as a common denominator: $\frac{11}{3} - \frac{9}{5} =: \frac{55}{15} - \frac{27}{15} = \frac{28}{15} = 1\frac{13}{15}$ (when rewritten to get rid of the partial fraction).

11. A: Dividing by 98 can be approximated by dividing by 100, which would mean shifting the decimal point of the numerator to the left by 2. The result is 4.2 and rounds to 4.

12. B: $4\frac{1}{3} + 3\frac{3}{4} = 4 + 3 + \frac{1}{3} + \frac{3}{4} = 7 + \frac{1}{3} + \frac{3}{4}$. Adding the fractions gives $\frac{1}{3} + \frac{3}{4} = \frac{4}{12} + \frac{9}{12} = \frac{13}{12} = 1 + \frac{1}{12}$. Thus, $7 + \frac{1}{3} + \frac{3}{4} = 7 + 1 + \frac{1}{12} = 8\frac{1}{12}$.

13. C: The average is calculated by adding all six numbers, then dividing by 6. The first five numbers have a sum of 25. If the total divided by 6 is equal to 6, then the total itself must be 36. The sixth number must be 36 – 25 = 11.

14. B: Multiplying by 10^{-3} means moving the decimal point three places to the left, putting in zeroes as necessary.

15. B: $\frac{5}{2} \div \frac{1}{3} = \frac{5}{2} \times \frac{3}{1} = \frac{15}{2} = 7.5$.

16. A: The total fraction taken up by green and red shirts will be $\frac{1}{3} + \frac{2}{5} = \frac{5}{15} + \frac{6}{15} = \frac{11}{15}$. The remaining fraction is $1 - \frac{11}{15} = \frac{15}{15} - \frac{11}{15} = \frac{4}{15}$.

17. C: If she has used 1/3 of the paint, she has 2/3 remaining. $2\frac{1}{2}$ gallons are the same as $\frac{5}{2}$ gallons. The calculation is $\frac{2}{3} \times \frac{5}{2} = \frac{5}{3} = 1\frac{2}{3}$ gallons.

18. D: 80%. To convert a fraction to a percent, the fraction is first converted to a decimal. To do so, the numerator is divided by the denominator: $4 \div 5 = 0.8$. To convert a decimal to a percent, the number is multiplied by 100: $0.8 \times 100 = 80\%$.

19. D: 1, 2, 3, 4, 6, 12. A given number divides evenly by each of its factors to produce an integer (no decimals). The number 5, 7, 8, 9, 10, 11 (and their opposites) do not divide evenly into 12. Therefore, these numbers are not factors.

20. D: This problem can be solved using basic arithmetic. Xavier starts with 20 apples, then gives his sister half, so 20 divided by 2.

$$\frac{20}{2} = 10$$

He then gives his neighbor 6, so 6 is subtracted from 10.

$$10 - 6 = 4$$

Lastly, he uses ¾ of his apples to make an apple pie, so to find remaining apples, the first step is to subtract ¾ from one and then multiply the difference by 4.

$$\left(1 - \frac{3}{4}\right) \times 4 = ?$$

$$\left(\frac{4}{4} - \frac{3}{4}\right) \times 4 = ?$$

$$\left(\frac{1}{4}\right) \times 4 = 1$$

21. A: 847.90. The hundredth place value is located two digits to the right of the decimal point (the digit 9). The digit to the right of the place value is examined to decide whether to round up or keep the digit. In this case, the digit 6 is 5 or greater so the hundredth place is rounded up. When rounding up, if the digit to be increased is a 9, the digit to its left is increased by one and the digit in the desired place value is made a zero. Therefore, the number is rounded to 847.90.

22. B: $\frac{11}{15}$. Fractions must have like denominators to be added. The least common multiple of the denominators 3 and 5 is found. The LCM is 15, so both fractions should be changed to equivalent fractions with a denominator of 15. To determine the numerator of the new fraction, the old numerator is multiplied by the same number by which the old denominator is multiplied to obtain the new denominator. For the fraction $\frac{1}{3}$, 3 multiplied by 5 will produce 15. Therefore, the numerator is multiplied by 5 to produce the new numerator $\left(\frac{1\times5}{3\times5} = \frac{5}{15}\right)$. For the fraction $\frac{2}{5}$, multiplying both the numerator and denominator by 3 produces $\frac{6}{15}$. When fractions have like denominators, they are added by adding the numerators and keeping the denominator the same: $\frac{5}{15} + \frac{6}{15} = \frac{11}{15}$.

23. C: $31/36$

Set up the problem and find a common denominator for both fractions.

$$\frac{5}{12} + \frac{4}{9}$$

Multiply each fraction across by 1 to convert to a common denominator.

$$\frac{5}{12} \times \frac{3}{3} + \frac{4}{9} \times \frac{4}{4}$$

Once over the same denominator, add across the top. The total is over the common denominator.

$$\frac{15 + 16}{36} = \frac{31}{36}$$

24. A: 52/27

Set up the division problem.

$$\frac{4}{13} \div \frac{27}{169}$$

Flip the second fraction and multiply.

$$\frac{4}{13} \times \frac{169}{27}$$

Simplify and reduce with cross multiplication.

$$\frac{4}{1} \times \frac{13}{27}$$

Multiply across the top and across the bottom to solve.

$$\frac{4 \times 13}{1 \times 27} = \frac{52}{27}$$

25. A: 3 3/5

Divide.

$$\begin{array}{r} 3 \\ 15\overline{)54} \\ -45 \\ \hline 9 \end{array}$$

The result is 3 9/15.

Reduce the remainder for the final answer.

3 3/5

26. A: Using the order of operations, multiplication and division are computed first from left to right. Multiplication is on the left; therefore, the teacher should perform multiplication first.

27. D: 59/7

The original number was 8 3/7. Multiply the denominator by the whole number portion. Add the numerator and put the total over the original denominator.

$$\frac{(8\ \times 7) + 3}{7} = \frac{59}{7}$$

28. B: 93/8

The original number was 11 5/8. Multiply the denominator by the whole number portion. Add the numerator and put the total over the original denominator.

$$\frac{(8\ \times 11)\ + 5}{8} = \frac{93}{8}$$

29. B: 8.1

To round 8.067 to the nearest tenths, use the digit in the hundredths.

6 in the hundredths is greater than 5, so round up in the tenths.

8.$\underline{0}$67

0 becomes a 1.

8.1

30. A: The place value to the right of the thousandth place, which would be the ten-thousandth place, is what gets used. The value in the thousandth place is 7. The number in the place value to its right is greater than 4, so the 7 gets bumped up to 8. Everything to its right turns to a zero, to get 245.2680. The zero is dropped because it is part of the decimal.

College-Level Math

Algebraic Operations

Polynomials

Algebraic expressions are built out of monomials. A *monomial* is a variable raised to some power multiplied by a constant: ax^n, where *a* is any constant and *n* is a whole number. A constant is also a monomial.

A *polynomial* is a sum of monomials. Examples of polynomials include $3x^4 + 2x^2 - x - 3$ and $\frac{4}{5}x^3$. The latter is also a monomial. If the highest power of *x* is 1, the polynomial is called *linear*. If the highest power of *x* is 2, it is called *quadratic*.

Simplifying Rational Algebraic Expressions

A *rational expression* is a ratio or fraction of two polynomials. An expression is in *lowest terms* when the numerator and denominator have no common factors. The rational expression $\frac{7}{4x+3}$ is in lowest terms because there are no common factors between the numerator and denominator. The rational expression $\frac{x^2+2x+1}{x^2-1}$ can be simplified to $\frac{(x+1)(x+1)}{(x-1)(x+1)} = \frac{x+1}{x-1}$ because there is a common factor of $x + 1$.

Factoring

Factors for polynomials are similar to factors for integers—they are numbers, variables, or polynomials that, when multiplied together, give a product equal to the polynomial in question. One polynomial is a factor of a second polynomial if the second polynomial can be obtained from the first by multiplying by a third polynomial.

$6x^6 + 13x^4 + 6x^2$ can be obtained by multiplying together $(3x^4 + 2x^2)(2x^2 + 3)$. This means $2x^2 + 3$ and $3x^4 + 2x^2$ are factors of $6x^6 + 13x^4 + 6x^2$.

In general, finding the factors of a polynomial can be tricky. However, there are a few types of polynomials that can be factored in a straightforward way.

If a certain monomial divides each term of a polynomial, it can be factored out:

$$x^2 + 2xy + y^2 = (x + y)^2$$

$$x^2 - 2xy + y^2 = (x - y)^2$$

$$x^2 - y^2 = (x + y)(x - y)$$

$$x^3 + y^3 = (x + y)(x^2 - xy + y^2)$$

$$x^3 - y^3 = (x - y)(x^2 + xy + y^2)$$

$$x^3 + 3x^2y + 3xy^2 + y^3 = (x + y)^3$$

$$x^3 - 3x^2y + 3xy^2 - y^3 = (x - y)^3$$

These rules can be used in many combinations with one another. For example, the expression $3x^3 - 24$ factors to $3(x^3 - 8) = 3(x - 2)(x^2 + 2x + 4)$.

When factoring polynomials, a good strategy is to multiply the factors to check the result.

Expanding Polynomials

Some polynomials may need to be expanded to identify the final solution—*polynomial expansion* means that parenthetical polynomials are multiplied out so that the parentheses no longer exist. The polynomials will be in the form $(a + b)^n$ where n is a whole number greater than 2. The expression can be simplified using the *distributive property*, which states that a number, variable, or polynomial that is multiplied by a polynomial in parentheses should be multiplied by each item in the parenthetical polynomial. Here's one example:

$$(a + b)^2 = (a + b)(a + b) = a^2 + ab + ab + b^2 = a^2 + 2ab + b^2$$

Here's another example to consider:

$$\begin{aligned}(a + b)^3 = (a + b)(a + b)(a + b) = (a^2 + ab + ab + b^2)(a + b) = (a^2 + 2ab + b^2)(a + b) \\ = a^3 + 2a^2b + ab^2 + a^2b + 2ab^2 + b^3 = a^3 + 3a^2b + 3ab^2 + b^3\end{aligned}$$

Properties of Exponents

Exponents are used in mathematics to express a number or variable multiplied by itself a certain number of times. For example, x^3 means *x* is multiplied by itself three times. In this expression, x is called the *base*, and 3 is the *exponent*. Exponents can be used in more complex problems when they contain fractions and negative numbers.

Fractional exponents can be explained by looking first at the inverse of exponents, which are *roots*. Given the expression x^2, the square root can be taken, $\sqrt{x^2}$, cancelling out the 2 and leaving *x* by itself, if *x* is positive. Cancellation occurs because $\sqrt{x}$ can be written with exponents, instead of roots, as $x^{\frac{1}{2}}$. The numerator of 1 is the exponent, and the denominator of 2 is called the root (which is why it's referred to as *square root*). Taking the square root of x^2 is the same as raising it to the $\frac{1}{2}$ power. Written out in mathematical form, it takes the following progression:

$$\sqrt{x^2} = (x^2)^{\frac{1}{2}} = x$$

From properties of exponents, $2 \cdot \frac{1}{2} = 1$ is the actual exponent of *x*. Another example can be seen with $x^{\frac{4}{7}}$. The variable *x,* raised to four-sevenths, is equal to the seventh root of *x* to the fourth power: $\sqrt[7]{x^4}$. In general,

$$x^{\frac{1}{n}} = \sqrt[n]{x}$$

and

$$x^{\frac{m}{n}} = \sqrt[n]{x^m}$$

Negative exponents also involve fractions. Whereas y^3 can also be rewritten as $\frac{y^3}{1}$, y^{-3} can be rewritten as $\frac{1}{y^3}$. A negative exponent means the exponential expression must be moved to the opposite spot in a fraction to make the exponent positive. If the negative appears in the numerator, it moves to the denominator. If the negative appears in the denominator, it is moved to the numerator. In general, $a^{-n} = \frac{1}{a^n}$, and a^{-n} and a^n are reciprocals.

Take, for example, the following expression:

$$\frac{a^{-4}b^2}{c^{-5}}$$

Since *a* is raised to the negative fourth power, it can be moved to the denominator. Since *c* is raised to the negative fifth power, it can be moved to the numerator. The *b* variable is raised to the positive second power, so it does not move.

The simplified expression is as follows:

$$\frac{b^2c^5}{a^4}$$

In mathematical expressions containing exponents and other operations, the order of operations must be followed. *PEMDAS* states that exponents are calculated after any parenthesis and grouping symbols, but before any multiplication, division, addition, and subtraction.

The Evaluation of Positive Rational Roots and Exponents

There are a few rules for working with exponents. For any numbers a, b, m, n, the following hold true:

$$a^1 = a$$

$$1^a = 1$$

$$a^0 = 1$$

$$a^m \times a^n = a^{m+n}$$

$$a^m \div a^n = a^{m-n}$$

$$(a^m)^n = a^{m \times n}$$

$$(a \times b)^m = a^m \times b^m$$

$$(a \div b)^m = a^m \div b^m$$

Any number, including a fraction, can be an exponent. The same rules apply.

Manipulating Roots and Exponents

A *root* is a different way to write an exponent when the exponent is the reciprocal of a whole number. We use the *radical* symbol to write this in the following way: $\sqrt[n]{a} = a^{\frac{1}{n}}$. This quantity is called the *n-th root* of *a*. The *n* is called the *index* of the radical.

Note that if the *n*-th root of *a* is multiplied by itself *n* times, the result will just be *a*. If no number *n* is written by the radical, it is assumed that *n* is 2: $\sqrt{5} = 5^{\frac{1}{2}}$. The special case of the 2nd root is called the *square root*, and the third root is called the *cube root*.

A *perfect square* is a whole number that is the square of another whole number. For example, sixteen and 64 are perfect squares because 16 is the square of 4, and 64 is the square of 8.

Solutions of Equations and Inequalities

Solving Linear and Quadratic Equations and Inequalities

An *equation* is an expression that uses an equals sign to demonstrate that two quantities are equal to one another, such as the expression $x^2 - x = -4x + 3$. *Solving* an equation means to find all possible values that *x* can take which make the equation true.

Given an equation where one side is a polynomial in the variable *x* and the other side is zero, the solutions are also called the *roots* or *zeros* of the equation.

To solve an equation, the equation needs to be modified to determine the solution. Starting with an equation $a = b$, the following are also true equations:

$$a + c = b + c$$

$$a - c = b - c$$

$$ac = bc$$

$$a/c = b/c \text{ (provided that } c \text{ is not 0)}$$

$$a^c = b^c$$

$$\sqrt{a} = \pm\sqrt{b}$$

The following rule is important to remember when solving equations:

If $ab = 0$, then $a = 0$ or $b = 0$.

Sometimes, instead of an equation, an *inequality* is used to indicate that one quantity is less than another (or greater than another). They may specify that the two quantities might also be equal to each other. If the quantities are not allowed to equal one another, the expression is a *strict inequality*. For example, $x + 3 \leq 5$ is an inequality, while $7 - 2x > 1$ is a strict inequality.

A *solution set* is a collection of all values of a variable that solve an equation or an inequality. For inequalities, this can be illustrated on a number line by shading in the part of the number line that satisfies the inequality. An open circle on the number line indicates that one gets arbitrarily close to that point, but cannot actually touch that point while remaining in the solution set. For example, to graph the solution set for the inequality $x > 2$, it would look like the following:

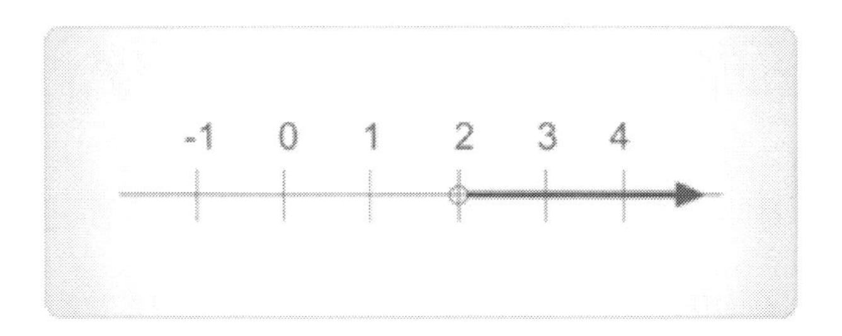

Solution Set x>2

Forms of Linear Equations

When graphing a linear function, note that the ratio of the change of the *y* coordinate to the change in the *x* coordinate is constant between any two points on the resulting line, no matter which two points are chosen. In other words, in a pair of points on a line, (x_1, y_1) and (x_2, y_2), with $x_1 \neq x_2$ so that the two points are distinct, then the ratio $\frac{y_2-y_1}{x_2-x_1}$ will be the same, regardless of which particular pair of points are chosen. This ratio, $\frac{y_2-y_1}{x_2-x_1}$, is called the *slope* of the line and is frequently denoted with the letter m. If slope *m* is positive, then the line goes upward when moving to the right, while if slope *m* is negative, then the line goes downward when moving to the right. If the slope is 0, then the line is called *horizontal*, and the *y* coordinate is constant along the entire line. In lines where the *x* coordinate is constant along the entire line, *y* is not actually a function of *x.* For such lines, the slope is not defined. These lines are called *vertical* lines.

Linear functions may take forms other than $y = ax + b$. The most common forms of linear equations are explained below:

- Standard Form: $\mathrm{Ax + By = C}$, in which the slope is given by $\mathrm{m} = \frac{-\mathrm{A}}{\mathrm{B}}$, and the y-intercept is given by $\frac{\mathrm{C}}{\mathrm{B}}$.
- Slope-Intercept Form: $\mathrm{y = mx + b}$, where the slope is m and the y intercept is b.
- Point-Slope Form: $\mathrm{y - y_1 = m(x - x_1)}$, where the slope is m and $(\mathrm{x_1, y_1})$ is any point on the chosen line.
- Two-Point Form: $\frac{\mathrm{y-y_1}}{\mathrm{x-x_1}} = \frac{\mathrm{y_2-y_1}}{\mathrm{x_2-x_1}}$, where $(\mathrm{x_1, y_1})$ and $(\mathrm{x_2, y_2})$ are any two distinct points on the chosen line. Note that the slope is given by $\mathrm{m} = \frac{\mathrm{y_2-y_1}}{\mathrm{x_2-x_1}}$.
- Intercept Form: $\frac{\mathrm{x}}{\mathrm{x_1}} + \frac{\mathrm{y}}{\mathrm{y_1}} = 1$, in which $\mathrm{x_1}$ is the x-intercept and $\mathrm{y_1}$ is the y-intercept.

These five ways to write linear equations are all useful in different circumstances. Depending on the given information, it may be easier to write one of the forms over another.

If $y = mx$, *y* is directly proportional to *x*. In this case, changing *x* by a factor changes *y* by that same factor. If $y = \frac{m}{x}$, *y* is inversely proportional to *x*. For example, if *x* is increased by a factor of 3, then *y* will be decreased by the same factor, 3.

Creating, Solving, or Interpreting a Linear Expression or Equation in One Variable

Linear expressions and equations are concise mathematical statements that can be written to model a variety of scenarios. Questions found pertaining to this topic will contain one variable only. A variable is an unknown quantity, usually denoted by a letter (*x, n, p,* etc.). In the case of linear expressions and equations, the power of the variable (its exponent) is 1. A variable without a visible exponent is raised to the first power.

Writing Linear Expressions and Equations

A linear expression is a statement about an unknown quantity expressed in mathematical symbols. The statement "five times a number added to forty" can be expressed as $5x + 40$. A linear equation is a statement in which two expressions (at least one containing a variable) are equal to each other. The statement "five times a number added to forty is equal to ten" can be expressed as $5x + 40 = 10$. Real-world scenarios can also be expressed mathematically. Consider the following:

> Bob had \$20 and Tom had \$4. After selling 4 ice cream cones to Bob, Tom has as much money as Bob.

The cost of an ice cream cone is an unknown quantity and can be represented by a variable. The amount of money Bob has after his purchase is four times the cost of an ice cream cone subtracted from his original \$20. The amount of money Tom has after his sale is four times the cost of an ice cream cone added to his original \$4. This can be expressed as: $20 - 4x = 4x + 4$, where *x* represents the cost of an ice cream cone.

When expressing a verbal or written statement mathematically, it is key to understand words or phrases that can be represented with symbols. The following are examples:

Symbol	Phrase
$+$	added to, increased by, sum of, more than
$-$	decreased by, difference between, less than, take away
x	multiplied by, 3 (4, 5 . . .) times as large, product of
$\div$	divided by, quotient of, half (third, etc.) of
$=$	is, the same as, results in, as much as
$x, t, n, etc.$	a number, unknown quantity, value of

Evaluating and Simplifying Algebraic Expressions

Given an algebraic expression, students may be asked to evaluate for given values of variable(s). In doing so, students will arrive at a numerical value as an answer. For example:

$$\text{Evaluate } a - 2b + ab\ for\ a = 3 \text{ and } b = -1$$

To evaluate an expression, the given values should be substituted for the variables and simplified using the order of operations. In this case: $(3) - 2(-1) + (3)(-1)$. Parentheses are used when substituting.

Given an algebraic expression, students may be asked to simplify the expression. For example:

Simplify $5x^2 - 10x + 2 - 8x^2 + x - 1$.

Simplifying algebraic expressions requires combining like terms. A term is a number, variable, or product of a number and variables separated by addition and subtraction. The terms in the above expressions are: $5x^2, -10x, 2, -8x^2, x$, and -1. Like terms have the same variables raised to the same powers (exponents). To combine like terms, the coefficients (numerical factor of the term including sign) are added, while the variables and their powers are kept the same. The example above simplifies to $-3x^2 - 9x + 1$.

Solving Linear Equations

When asked to solve a linear equation, it requires determining a numerical value for the unknown variable. Given a linear equation involving addition, subtraction, multiplication, and division, isolation of the variable is done by working backward. Addition and subtraction are inverse operations, as are multiplication and division; therefore, they can be used to cancel each other out.

The first steps to solving linear equations are to distribute if necessary and combine any like terms that are on the same side of the equation. Sides of an equation are separated by an $=$ sign. Next, the equation should be manipulated to get the variable on one side. Whatever is done to one side of an equation, must be done to the other side to remain equal. Then, the variable should be isolated by using inverse operations to undo the order of operations backward. Undo addition and subtraction, then undo multiplication and division. For example:

Solve $4(t - 2) + 2t - 4 = 2(9 - 2t)$

Distribute: $4t - 8 + 2t - 4 = 18 - 4t$

Combine like terms: $6t - 12 = 18 - 4t$

Add 4t to each side to move the variable: $10t - 12 = 18$

Add 12 to each side to isolate the variable: $10t = 30$

Divide each side by 10 to isolate the variable: $t = 3$

The answer can be checked by substituting the value for the variable into the original equation and ensuring both sides calculate to be equal.

Creating, Solving, or Interpreting Linear Inequalities in One Variable

Linear inequalities and linear equations are both comparisons of two algebraic expressions. However, unlike equations in which the expressions are equal to each other, linear inequalities compare expressions that are unequal. Linear equations typically have one value for the variable that makes the statement true. Linear inequalities generally have an infinite number of values that make the statement true. Exceptions to these last two statements are covered in Section 6.

Writing Linear Inequalities

Linear inequalities are a concise mathematical way to express the relationship between unequal values. More specifically, they describe in what way the values are unequal. A value could be greater than ($>$); less than ($<$); greater than or equal to ($\geq$); or less than or equal to ($\leq$) another value. The statement "five times a number added to forty is more than sixty-five" can be expressed as $5x + 40 > 65$. Common words and phrases that express inequalities are:

Symbol	**Phrase**
$<$	is under, is below, smaller than, beneath
$>$	is above, is over, bigger than, exceeds
$\leq$	no more than, at most, maximum
$\geq$	no less than, at least, minimum

Solving Linear Inequalities

When solving a linear inequality, the solution is the set of all numbers that makes the statement true. The inequality $x + 2 \geq 6$ has a solution set of 4 and every number greater than 4 (4.0001, 5, 12, 107, etc.). Adding 2 to 4 or any number greater than 4 would result in a value that is greater than or equal to 6. Therefore, $x \geq 4$ would be the solution set.

Solution sets for linear inequalities often will be displayed using a number line. If a value is included in the set ($\geq$ or $\leq$), there is a shaded dot placed on that value and an arrow extending in the direction of the solutions. For a variable $>$ or $\geq$ a number, the arrow would point right on the number line (the direction where the numbers increase); and if a variable is $<$ or $\leq$ a number, the arrow would point left (where the numbers decrease). If the value is not included in the set ($>$ or $<$), an open circle on that value would be used with an arrow in the appropriate direction.

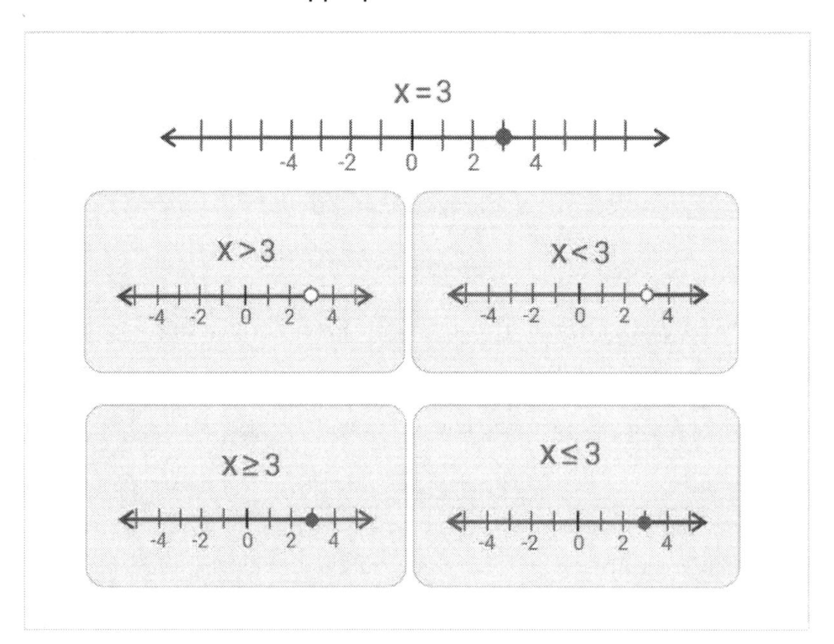

Students may be asked to write a linear inequality given a graph of its solution set. To do so, they should identify whether the value is included (shaded dot or open circle) and the direction in which the arrow is pointing.

In order to algebraically solve a linear inequality, the same steps should be followed as in solving a linear equation (see section on *Solving Linear Equations*). The inequality symbol stays the same for all operations EXCEPT when dividing by a negative number. If dividing by a negative number while solving an inequality, the relationship reverses (the sign flips). Dividing by a positive does not change the relationship, so the sign stays the same. In other words, $>$ switches to $<$ and vice versa. An example is shown below.

Solve $-2(x+4) \leq 22$

Distribute: $-2x - 8 \leq 22$

Add 8 to both sides: $-2x \leq 30$

Divide both sides by -2: $x \geq 15$

Equation Systems

Sometimes, a problem involves multiple variables and multiple equations that must all hold true at the same time. There are a few basic rules to keep in mind when solving systems of equations.

A single equation can be changed by doing the same operation to both sides, just as with one equation.

If one of the equations gives an expression for one of the variables in terms of other variables and constants, the expression can be substituted into the other equation, in place of the variable. This means the other equations will have one less variable in them.

If two equations are in the form of $a = b \text{ and } c = d$, then a new equation can be formed by adding the equations together, $a + c = b + d$, or subtracting the equations, $a - c = b - d$. This can eliminate one of the variables from an equation.

The general approach is to find a way to change one of the equations so that one variable is isolated, and then substitute that value (or expression) for the variable into the other equations.

The simplest case is a *linear system of two equations*, which has the form $ax + by = c, dx + ey = f$.

To solve linear systems of equations, use the same process to solve one equation in order to isolate one of the variables. Here's an example, using the linear system of equations:

$$2x - 3y = 2, 4x + 4y = 3$$

The first equation is multiplied on both sides by -2, which gives $-4x + 6y = -4$.

Adding this equation to the second equation will allow cancellation of the *x* term: $4x + 4y - 4x + 6y = 3 - 4$.

The result can be simplified to get $10y = -1$, which simplifies to $y = -\frac{1}{10}$.

The solution can be substituted into either of the original equations to find a value for *x*. Using the first equation, $2x - 3\left(-\frac{1}{10}\right) = 2$.

This simplifies to $2x + \frac{3}{10} = 2$, then to $2x = \frac{17}{10}$, and finally $x = \frac{17}{20}$.

The final solution is $x = \frac{17}{20}, y = -\frac{1}{10}$.

To check the validity of the answer, both solutions can be substituted into either original equation, which should result in a true statement.

An alternative way to solve this system would be to solve the first equation to get an expression for *y* in terms of *x*.

Subtracting $2x$ from both sides results in $-3y = 2 - 2x$.

Dividing both sides by -3 would be $y = \frac{2}{3}x - \frac{2}{3}$.

Then, this expression can be substituted into the second equation, getting $4x + 4\left(\frac{2}{3}x - \frac{2}{3}\right) = 3$.

This only involves the variable *x*, which can now be solved. Once the value for *x* is obtained, it can be substituted into either equation to solve for *y*.

There is one important issue to note here. If one of the equations in the system can be made to look identical to another equation, then it is *redundant*. The set of solutions will then be all pairs that satisfy the other equation.

For instance, in the system of equations, $2x - y = 1, -4x + 2y = -2$, the second equation can be made into the first equation by dividing both sides by -2. Thus, the solution set will be all pairs satisfying $2x - y = 1$, which simplifies to $y = 2x - 1$.

For a pair of linear equations, the simplest way to see if one equation is redundant is to rewrite each equation to the form $ax + by = c$. If one equation can be obtained from the other in this form by multiplying both sides by some constant, then the equations are redundant, and the answer to the system would be all real numbers.

It is also possible for the two equations to be *inconsistent*, which occurs when the two equations can be made into the form $ax + by = c, ax + by = d$, with *c* and *d* being different numbers. The two equations are inconsistent if, while trying to solve them, it is determined that an equation is false, such as $3 = 2$. This result shows that there are no solutions to that system of equations.

For linear systems of two equations with two variables, there will always be a single solution unless one of the two equations is redundant or the equations are inconsistent, in which case there are no solutions.

Other Algebraic Functions

A *function* $f(x)$ is a mathematical object which takes one number, *x*, as an input and gives a number in return. The input is called the *independent variable*. If the variable is set equal to the output, as in $y = f(x)$, then this is called the *dependent variable*. To indicate the dependent value a function, y, gives for a specific independent variable, x, the notation y = $f(x)$ is used.

The *domain* of a function is the set of values that the independent variable is allowed to take. Unless otherwise specified, the domain is any value for which the function is well defined. The *range* of the function is the set of possible outputs for the function.

In many cases, a function can be defined by giving an equation. For instance, $f(x) = x^2$ indicates that given a value for *x*, the output of *f* is found by squaring *x*.

Not all equations in *x* and *y* can be written in the form $y = f(x)$. An equation can be written in such a form if it satisfies the *vertical line test*: no vertical line meets the graph of the equation at more than a single point. In this case, *y* is said to be a *function of x*. If a vertical line meets the graph in two places, then this equation cannot be written in the form $y = f(x)$.

The graph of a function $f(x)$ is the graph of the equation $y = f(x)$. Thus, it is the set of all pairs (x, y) where $y = f(x)$. In other words, it is all pairs $\left(x, f(x)\right)$. The *x*-intercepts are called the *zeros* of the function. The *y*-intercept is given by $f(0)$.

If, for a given function f, the only way to get $f(a) = f(b)$ is for $a = b$, then *f* is *one-to-one*. Often, even if a function is not one-to-one on its entire domain, it is one-to-one by considering a restricted portion of the domain.

A function $f(x) = k$ for some number *k* is called a *constant function*. The graph of a constant function is a horizontal line.

The function $f(x) = x$ is called the *identity function*. The graph of the identity function is the diagonal line pointing to the upper right at 45 degrees, $y = x$.

Given two functions, $f(x) \;\; and \;\; g(x)$, new functions can be formed by adding, subtracting, multiplying, or dividing the functions. Any algebraic combination of the two functions can be performed, including one function being the exponent of the other. If there are expressions for *f* and *g*, then the result can be found by performing the desired operation between the expressions. So, if $f(x) = x^2$ and $g(x) = 3x$, then $f \cdot g(x) = x^2 \cdot 3x = 3x^3$.

Given two functions, $f(x) \; and \;\; g(x)$, where the domain of *g* contains the range of *f*, the two functions can be combined together in a process called *composition*. The function—"*g* composed of *f*"—is written $(g \circ f)(x) = g(f(x))$. This requires the input of x into *f*, then taking that result and plugging it in to the function *g*.

If *f* is one-to-one, then there is also the option to find the function $f^{-1}(x)$, called the *inverse* of *f*. Algebraically, the inverse function can be found by writing y in place of $f(x)$, and then solving for *x*. The inverse function also makes this statement true: $f^{-1}\left(f(x)\right) = x$.

Computing the inverse of a function *f* entails the following procedure:

Given $f(x) = x^2$, with a domain of $x \geq 0$

$x = y^2$is written down to find the inverse

The square root of both sides is determined to solve for *y*

Normally, this would mean $\pm\sqrt{x} = y$. However, the domain of *f* does not include the negative numbers, so the negative option needs to be eliminated.

The result is $y = \sqrt{x}$, so $f^{-1}(x) = \sqrt{x}$, with a domain of $x \geq 0$.

A function is called *monotone* if it is either always increasing or always decreasing. For example, the functions $f(x) = 3x$ and $f(x) = -x^5$ are monotone.

An *even function* looks the same when flipped over the *y*-axis: $f(x) = f(-x)$. The following image shows a graphic representation of an even function.

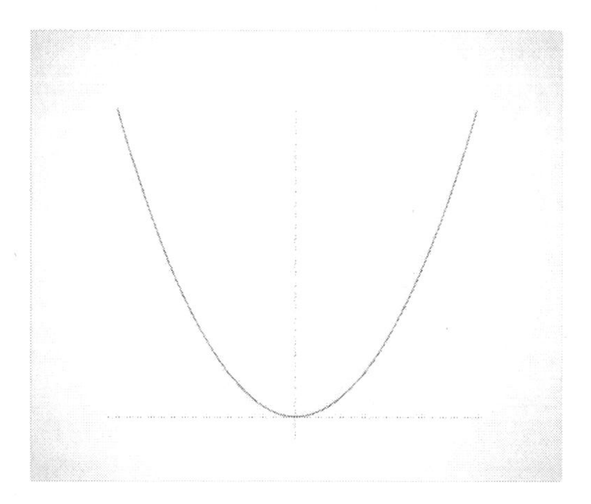

An *odd function* looks the same when flipped over the *y*-axis and then flipped over the *x*-axis: $f(x) = -f(-x)$. The following image shows an example of an odd function.

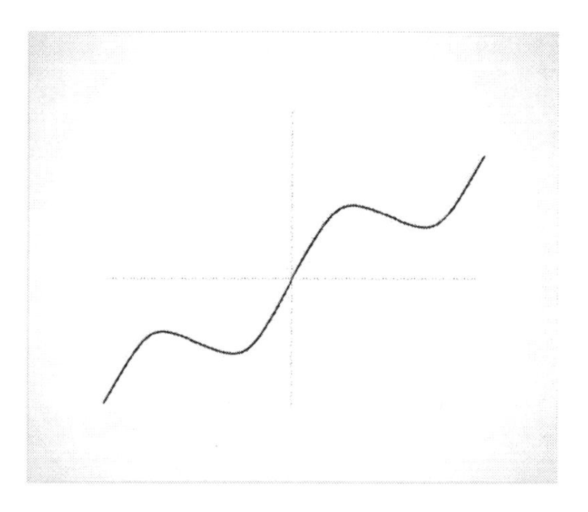

Coordinate Geometry

Plane Geometry

Algebraic equations can be used to describe geometric figures in the plane. The method for doing so is to use the *Cartesian coordinate plane*. The idea behind these Cartesian coordinates (named for mathematician and philosopher Descartes) is that from a specific point on the plane, known as the *center*, one can specify any other point by saying *how far to the right or left* and *how far up or down*.

The plane is covered with a grid. The two directions, right to left and bottom to top, are called *axes* (singular *axis*). When working with *x* and *y* variables, the *x* variable corresponds to the right and left axis, and the *y* variable corresponds to the up and down axis.

Any point on the grid is found by specifying how far to travel from the center along the *x*-axis and how far to travel along the *y*-axis. The ordered pair can be written as (x, y). A positive *x* value means go to the right on the *x*-axis, while a negative *x* value means to go to the left. A positive *y* value means to go up, while a negative value means to go down. Several points are shown as examples in the figure.

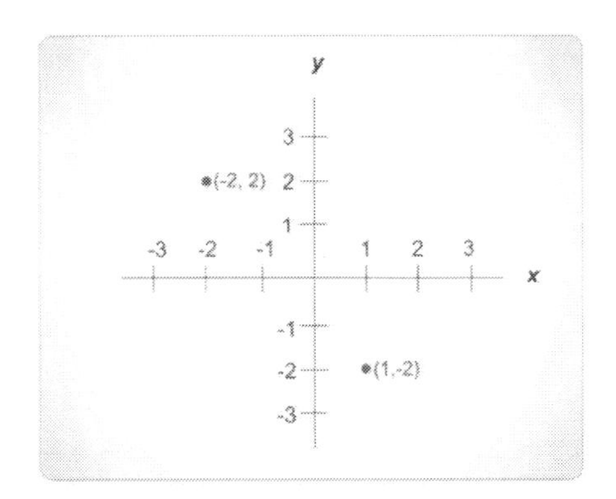

Cartesian Coordinate Plane

The Coordinate Plane

The coordinate plane can be divided into four *quadrants*. The upper-right part of the plane is called the *first quadrant*, where both x and y are positive. The *second quadrant* is the upper-left, where x is negative but y is positive. The *third quadrant* is the lower left, where both x and y are negative. Finally, the *fourth quadrant* is in the lower right, where x is positive but y is negative. These quadrants are often written with Roman numerals:

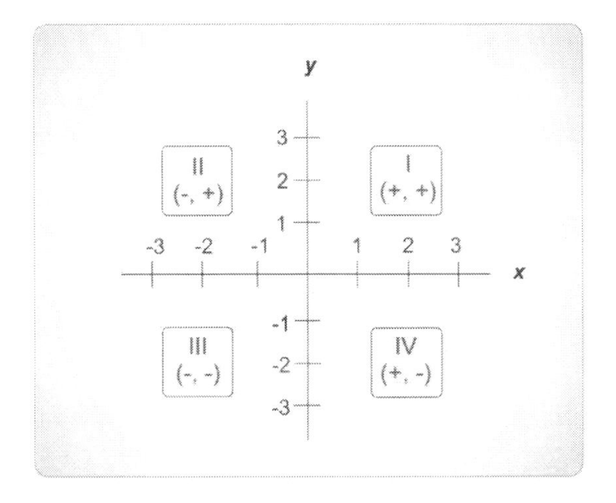

In addition to graphing individual points as shown above, the graph lines and curves in the plane can be graphed corresponding to equations. In general, if there is some equation involving x and y, then the *graph* of that equation consists of all the points (x, y) in the Cartesian coordinate plane, which satisfy this equation.

Given the equation $y = x + 2$, the point $(0, 2)$ is in the graph, since $2 = 0 + 2$ is a true equation. However, the point $(1, 4)$ will *not* be in the graph, because $4 = 1 + 2$ is false.

Straight Lines

The simplest equations to graph are the equations whose graphs are lines, called *linear equations*. Every linear equation can be rewritten algebraically so that it looks like $Ax + By = C$.

First, the ratio of the change in the y coordinate to the change in the x coordinate is constant for any two distinct points on the line. In any pair of points on a line, two points, (x_1, y_1) and (x_2, y_2)—

where $x_1 \neq x_2$—the ratio $\frac{y_2 - y_1}{x_2 - x_1}$ will always be the same, even if another pair of points is used.

This ratio, $\frac{y_2 - y_1}{x_2 - x_1}$, is called the *slope* of the line and is often denoted with the letter m. If the slope is *positive*, then the line goes upward when moving to the right. If the slope is *negative*, then it moves downward when moving to the right. If the slope is 0, then the line is *horizontal*, and the y coordinate is constant along the entire line. For lines where the x coordinate is constant along the entire line, the slope is not defined, and these lines are called *vertical* lines.

The y coordinate of the point where the line touches the y-axis is called the *y-intercept* of the line. It is often denoted by the letter b, used in the form of the linear equation $y = mx + b$. The x coordinate of the point where the line touches the x-axis is called the *x-intercept*. It is also called the *zero* of the line.

Suppose two lines have slopes m_1 and m_2. If the slopes are equal, $m_1 = m_2$, then the lines are *parallel*. Parallel lines never meet one another. If $m_1 = -\frac{1}{m_2}$, then the lines are called *perpendicular* or *orthogonal*. Their slopes can also be called opposite reciprocals of each other.

There are several convenient ways to write down linear equations. The common forms are listed here:

Standard Form: $Ax + By = C$, where the slope is given by $\frac{-A}{B}$, and the y-intercept is given by $\frac{C}{B}$.

Slope-Intercept Form: $y = mx + b$, where the slope is *m*, and the y-intercept is *b*.

Point-Slope Form: $y - y_1 = m(x - x_1)$, where *m* is the slope, and (x_1, y_1) is any point on the line.

Two-Point Form: $\frac{y - y_1}{x - x_1} = \frac{y_2 - y_1}{x_2 - x_1}$, where (x_1, y_1), and (x_2, y_2) are any two distinct points on the line.

Intercept Form: $\frac{x}{x_1} + \frac{y}{y_1} = 1$, where x_1 is the x-intercept, and y_1 is the y-intercept.

Depending upon the given information, different forms of the linear equation can be easier to write down than others. When given two points, the two-point form is easy to write down. If the slope and a single point is known, the point-slope form is easiest to start with. In general, which form to start with depends upon the given information.

Conics

The graph of an equation of the form $y = ax^2 + bx + c$ or $x = ay^2 + by + c$ is called a *parabola*.

The graph of an equation of the form $\frac{x^2}{a^2} - \frac{y^2}{b^2} = 1$ or $-\frac{x^2}{a^2} + \frac{y^2}{b^2} = 1$ is called a *hyperbola*.

The graph of an equation of the form $\frac{(x-x_0)^2}{a^2} + \frac{(y-y_0)^2}{b^2} = 1$ is called an *ellipse*. If $a = b$ then this is a circle with *radius* $r = \frac{1}{a}$.

Sets of Points in the Plane

The *midpoint* between two points, (x_1, y_1) and (x_2, y_2), is given by taking the average of the *x* coordinates and the average of the *y* coordinates: $\left(\frac{x_1+x_2}{2}, \frac{y_1+y_2}{2}\right)$.

The *distance* between two points, (x_1, y_1) and (x_2, y_2), is given by the *Pythagorean formula*, $\sqrt{(x_2 - x_1)^2 + (y_2 - y_1)^2}$.

To find the perpendicular distance between a line $Ax + By = C$ and a point (x_1, y_1) not on the line, we need to use the formula $\frac{|Ax_1+By_1+C|}{\sqrt{A^2+B^2}}$.

Transformations of a Plane

Given a figure drawn on a plane, many changes can be made to that figure, including *rotation*, *translation*, and *reflection*. Rotations turn the figure about a point, translations slide the figure, and reflections flip the figure over a specified line. When performing these transformations, the original figure is called the *pre-image*, and the figure after transformation is called the *image*.

More specifically, *translation* means that all points in the figure are moved in the same direction by the same distance. In other words, the figure is slid in some fixed direction. Of course, while the entire figure is slid by the same distance, this does not change any of the measurements of the figures involved. The result will have the same distances and angles as the original figure.

In terms of Cartesian coordinates, a translation means a shift of each of the original points (x, y) by a fixed amount in the *x* and *y* directions, to become $(x + a, y + b)$.

Another procedure that can be performed is called *reflection*. To do this, a line in the plane is specified, called the *line of reflection*. Then, take each point and flip it over the line so that it is the same distance from the line but on the opposite side of it. This does not change any of the distances or angles involved, but it does reverse the order in which everything appears.

To reflect something over the *x*-axis, the points (x, y) are sent to $(x, -y)$. To reflect something over the *y*-axis, the points (x, y) are sent to the points $(-x, y)$. Flipping over other lines is not something easy to express in Cartesian coordinates. However, by drawing the figure and the line of reflection, the distance to the line and the original points can be used to find the reflected figure.

Example: Reflect this triangle with vertices (-1, 0), (2, 1), and (2, 0) over the *y*-axis. The pre-image is shown below.

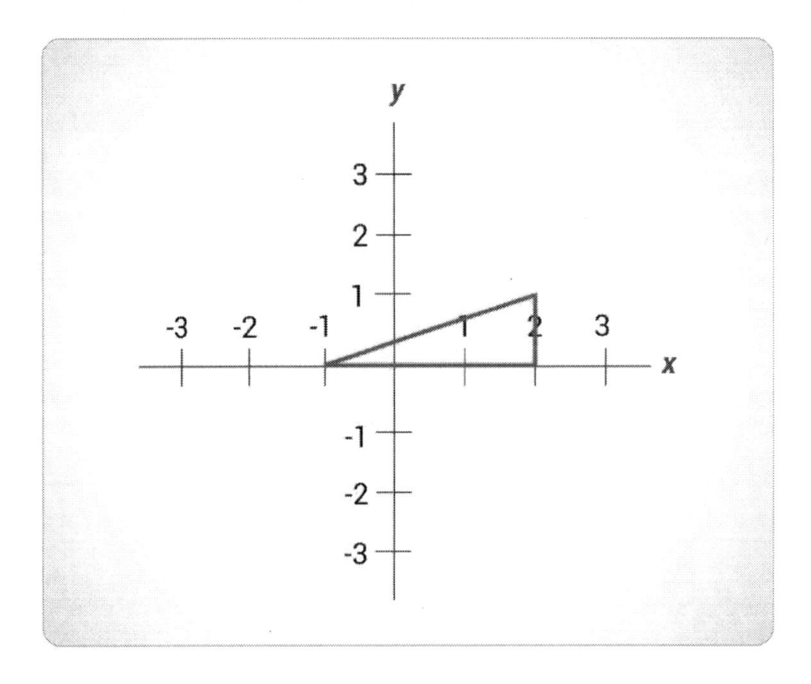

To do this, flip the *x* values of the points involved to the negatives of themselves, while keeping the *y* values the same. The image is shown here.

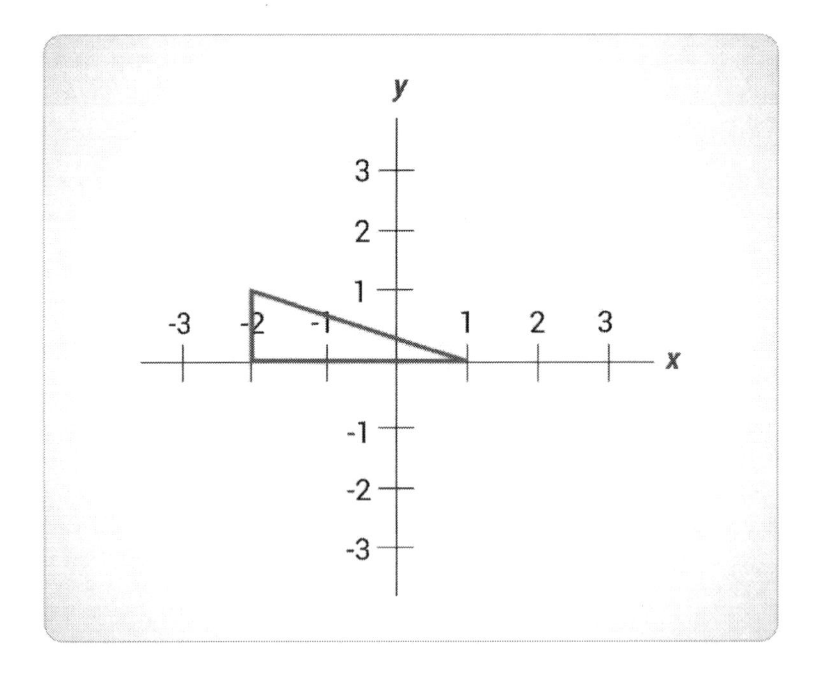

The new vertices will be (1, 0), (-2, 1), and (-2, 0).

Another procedure that does not change the distances and angles in a figure is *rotation*. In this procedure, pick a center point, then rotate every vertex along a circle around that point by the same angle. This procedure is also not easy to express in Cartesian coordinates, and this is not a requirement on this test. However, as with reflections, it's helpful to draw the figures and see what the result of the rotation would look like. This transformation can be performed using a compass and protractor.

Each one of these transformations can be performed on the coordinate plane without changes to the original dimensions or angles.

If two figures in the plane involve the same distances and angles, they are called *congruent figures*. In other words, two figures are congruent when they go from one form to another through reflection, rotation, and translation, or a combination of these.

Remember that rotation and translation will give back a new figure that is identical to the original figure, but reflection will give back a mirror image of it.

To recognize that a figure has undergone a rotation, check to see that the figure has not been changed into a mirror image, but that its orientation has changed (that is, whether the parts of the figure now form different angles with the *x* and *y* axes).

To recognize that a figure has undergone a translation, check to see that the figure has not been changed into a mirror image, and that the orientation remains the same.

To recognize that a figure has undergone a reflection, check to see that the new figure is a mirror image of the old figure.

Keep in mind that sometimes a combination of translations, reflections, and rotations may be performed on a figure.

Dilation

A *dilation* is a transformation that preserves angles, but not distances. This can be thought of as stretching or shrinking a figure. If a dilation makes figures larger, it is called an *enlargement*. If a dilation makes figures smaller, it is called a *reduction*. The easiest example is to dilate around the origin. In this case, multiply the *x* and *y* coordinates by a *scale factor*, k, sending points (x, y) to (kx, ky).

As an example, draw a dilation of the following triangle, whose vertices will be the points (-1, 0), (1, 0), and (1, 1).

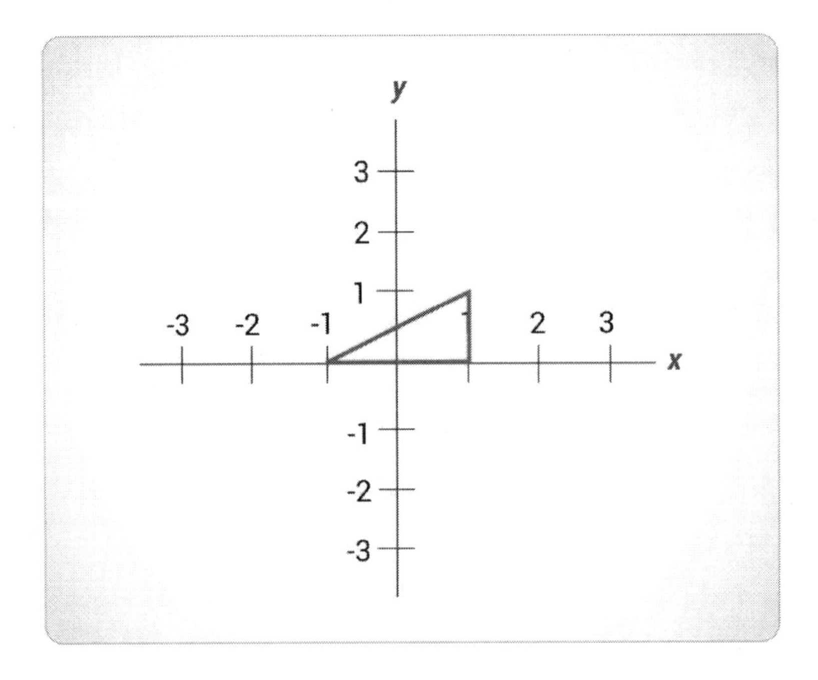

For this problem, dilate by a scale factor of 2, so the new vertices will be (-2, 0), (2, 0), and (2, 2).

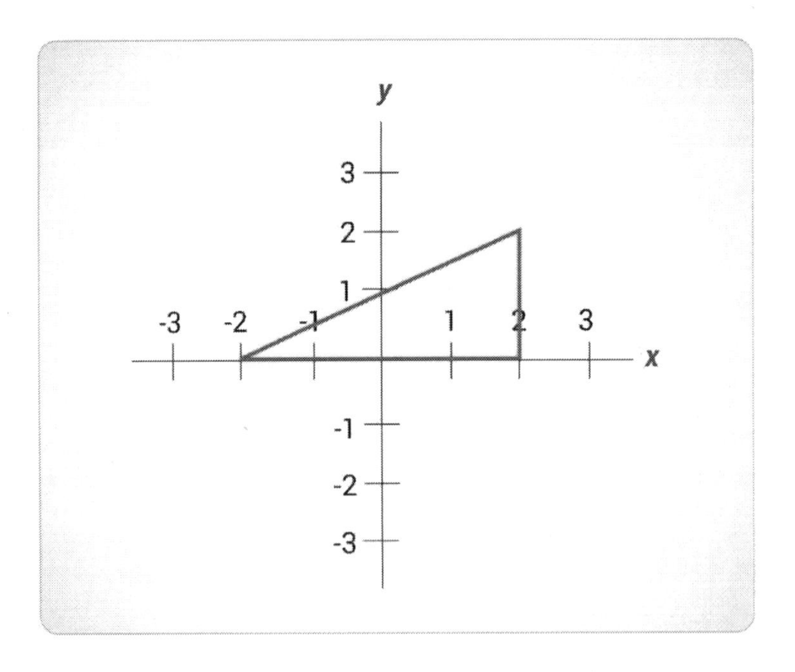

Note that after a dilation, the distances between the vertices of the figure will have changed, but the angles remain the same. The two figures that are obtained by dilation, along with possibly translation, rotation, and reflection, are all *similar* to one another. Another way to think of this is that similar figures have the same number of vertices and edges, and their angles are all the same. Similar figures have the same basic shape, but are different in size.

Symmetry

Using the types of transformations above, if an object can undergo these changes and not appear to have changed, then the figure is symmetrical. If an object can be split in half by a line and flipped over that line to lie directly on top of itself, it is said to have *line symmetry*. An example of both types of figures is seen below.

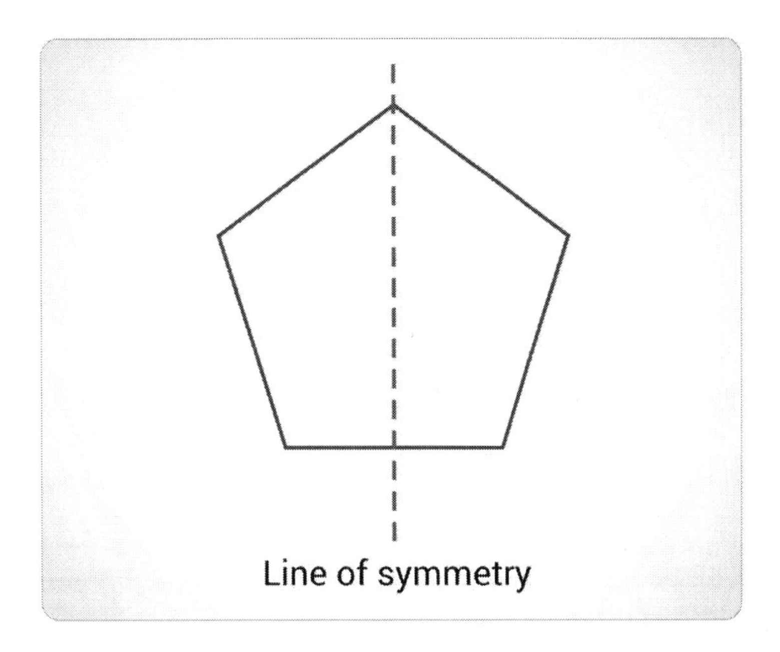

If an object can be rotated about its center to any degree smaller than 360, and it lies directly on top of itself, the object is said to have *rotational symmetry*. An example of this type of symmetry is shown below. The pentagon has an order of 5.

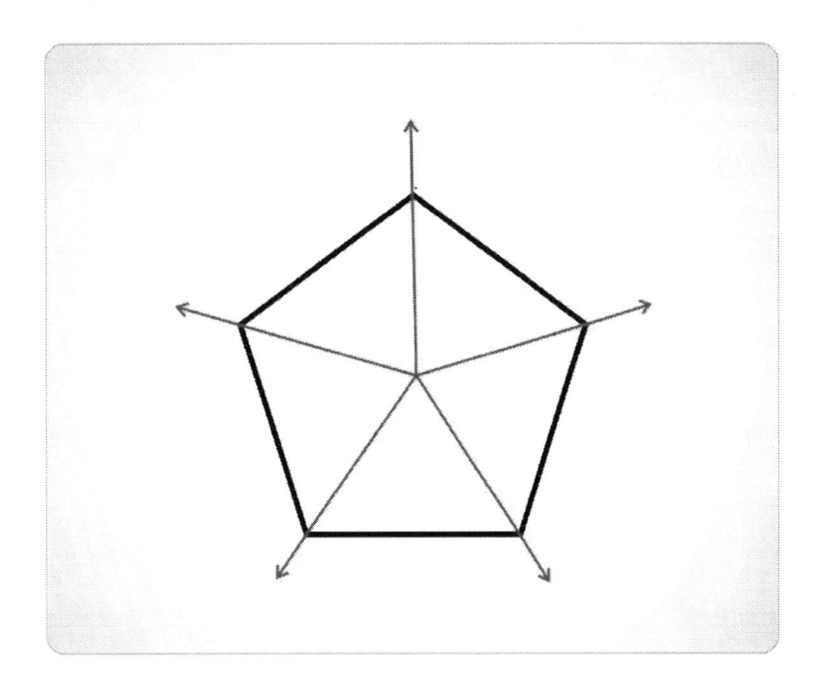

The rotational symmetry lines in the figure above can be used to find the angles formed at the center of the pentagon. Knowing that all of the angles together form a full circle, at 360 degrees, the figure can be split into 5 angles equally. By dividing the 360° by 5, each angle is 72°.

Given the length of one side of the figure, the perimeter of the pentagon can also be found using rotational symmetry. If one side length was 3 cm, that side length can be rotated onto each other side length four times. This would give a total of 5 side lengths equal to 3 cm. To find the perimeter, or distance around the figure, multiply 3 by 5. The perimeter of the figure would be 15 cm.

If a line cannot be drawn anywhere on the object to flip the figure onto itself or rotated less than or equal to 180 degrees to lay on top of itself, the object is asymmetrical. Examples of these types of figures are shown below.

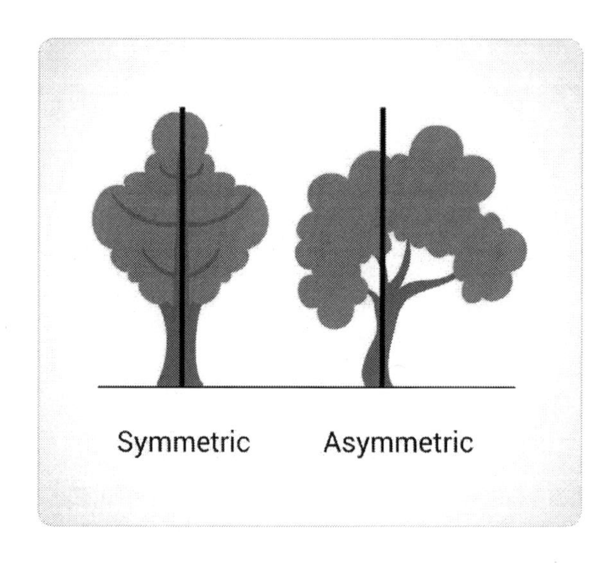

Similar Figures and Proportions

Sometimes, two figures are similar, meaning they have the same basic shape and the same interior angles, but they have different dimensions. If the ratio of two corresponding sides is known, then that ratio, or scale factor, holds true for all of the dimensions of the new figure.

Here is an example of applying this principle. Suppose that Lara is 5 feet tall and is standing 30 feet from the base of a light pole, and her shadow is 6 feet long. How high is the light on the pole? To figure this, it helps to make a sketch of the situation:

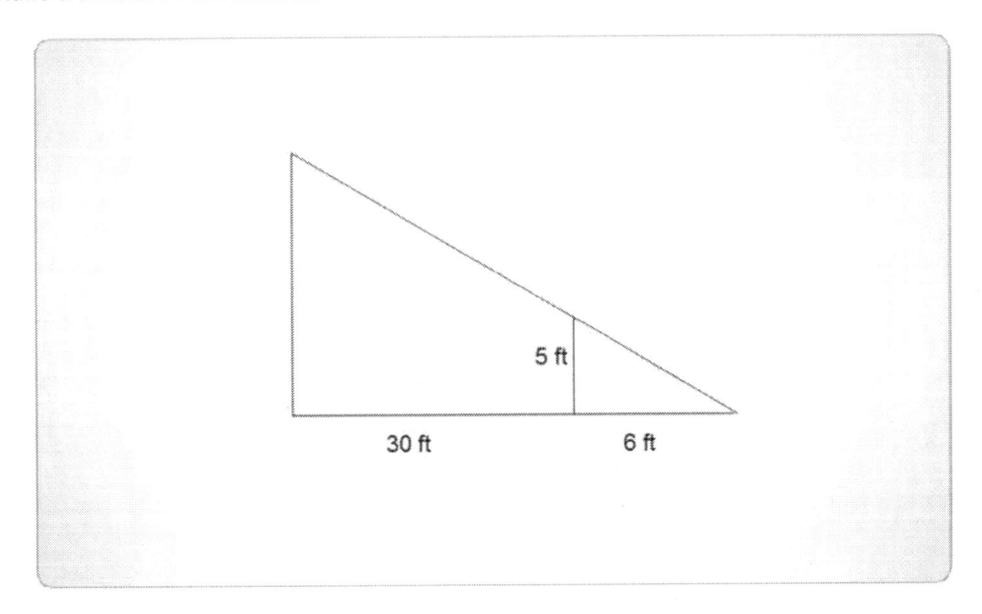

The light pole is the left side of the triangle. Lara is the 5-foot vertical line. Notice that there are two right triangles here, and that they have all the same angles as one another. Therefore, they form similar triangles. So, figure the ratio of proportionality between them.

The bases of these triangles are known. The small triangle, formed by Lara and her shadow, has a base of 6 feet. The large triangle, formed by the light pole along with the line from the base of the pole out to the end of Lara's shadow is $30 + 6 = 36$ feet long. So, the ratio of the big triangle to the little triangle will be $\frac{36}{6} = 6$. The height of the little triangle is 5 feet. Therefore, the height of the big triangle will be $6 \cdot 5 = 30$ feet, meaning that the light is 30 feet up the pole.

Notice that the perimeter of a figure changes by the ratio of proportionality between two similar figures, but the area changes by the *square* of the ratio. This is because if the length of one side is doubled, the area is quadrupled.

As an example, suppose two rectangles are similar, but the edges of the second rectangle are three times longer than the edges of the first rectangle. The area of the first rectangle is 10 square inches. How much more area does the second rectangle have than the first?

To answer this, note that the area of the second rectangle is $3^2 = 9$ times the area of the first rectangle, which is 10 square inches. Therefore, the area of the second rectangle is going to be $9 \cdot 10 = 90$ square inches. This means it has $90 - 10 = 80$ square inches more area than the first rectangle.

As a second example, suppose *X* and *Y* are similar right triangles. The hypotenuse of *X* is 4 inches. The area of *Y* is $\frac{1}{4}$ the area of *X*. What is the hypotenuse of *Y*?

First, realize the area has changed by a factor of $\frac{1}{4}$. The area changes by a factor that is the *square* of the ratio of changes in lengths, so the ratio of the lengths is the square root of the ratio of areas. That means that the ratio of lengths must be is $\sqrt{\frac{1}{4}} = \frac{1}{2}$, and the hypotenuse of *Y* must be $\frac{1}{2} \cdot 4 = 2$ inches.

Volumes between similar solids change like the cube of the change in the lengths of their edges. Likewise, if the ratio of the volumes between similar solids is known, the ratio between their lengths is known by finding the cube root of the ratio of their volumes.

For example, suppose there are two similar rectangular pyramids *X* and *Y*. The base of *X* is 1 inch by 2 inches, and the volume of *X* is 8 inches. The volume of *Y* is 64 inches. What are the dimensions of the base of *Y*?

To answer this, first find the ratio of the volume of *Y* to the volume of *X*. This will be given by $\frac{64}{8} = 8$. Now the ratio of lengths is the cube root of the ratio of volumes, or $\sqrt[3]{8} = 2$. So, the dimensions of the base of *Y* must be 2 inches by 4 inches.

Volumes and Surface Areas

Geometry in three dimensions is similar to geometry in two dimensions. The main new feature is that three points now define a unique *plane* that passes through each of them. Three dimensional objects can be made by putting together two dimensional figures in different surfaces. Below, some of the possible three dimensional figures will be provided, along with formulas for their volumes and surface areas.

A rectangular prism is a box whose sides are all rectangles meeting at 90° angles. Such a box has three dimensions: length, width, and height. If the length is *x*, the width is *y*, and the height is *z*, then the volume is given by $V = xyz$.

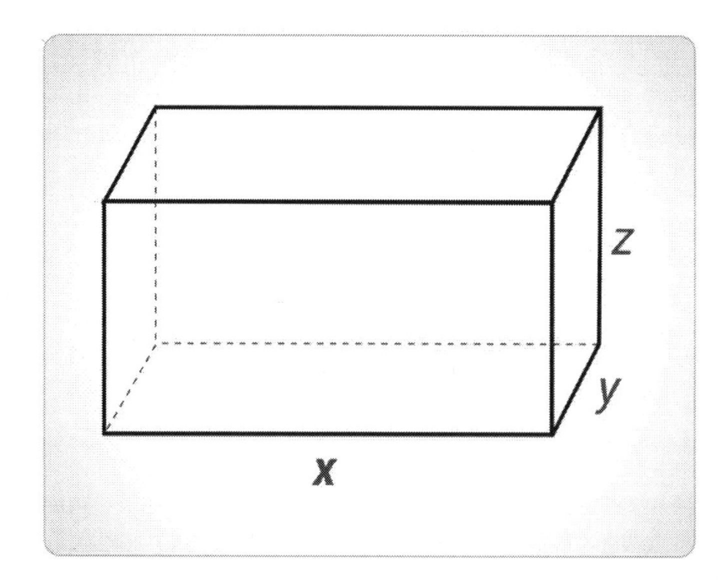

The surface area will be given by computing the surface area of each rectangle and adding them together. There are a total of six rectangles. Two of them have sides of length *x* and *y*, two have sides of length y and z, and two have sides of length x and z. Therefore, the total surface area will be given by $SA = 2xy + 2yz + 2xz$.

A *rectangular pyramid* is a figure with a rectangular base and four triangular sides that meet at a single vertex. If the rectangle has sides of length x and y, then the volume will be given by $V = \frac{1}{3}xyh$.

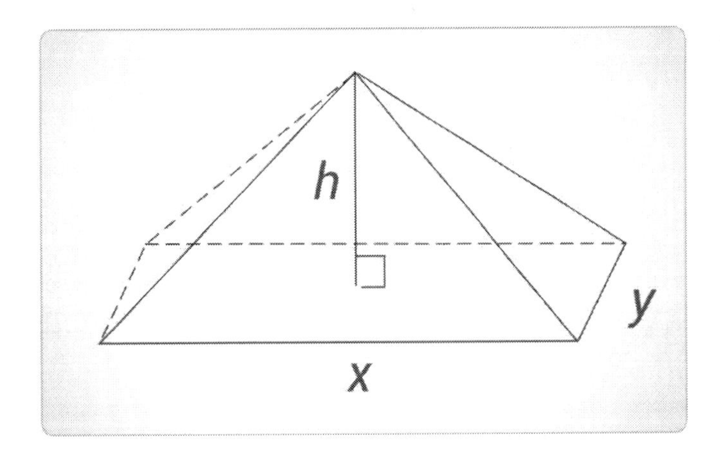

To find the surface area, the dimensions of each triangle need to be known. However, these dimensions can differ depending on the problem in question. Therefore, there is no general formula for calculating total surface area.

A *sphere* is a set of points all of which are equidistant from some central point. It is like a circle, but in three dimensions. The volume of a sphere of radius r is given by $V = \frac{4}{3}\pi r^3$. The surface area is given by $A = 4\pi r^2$.

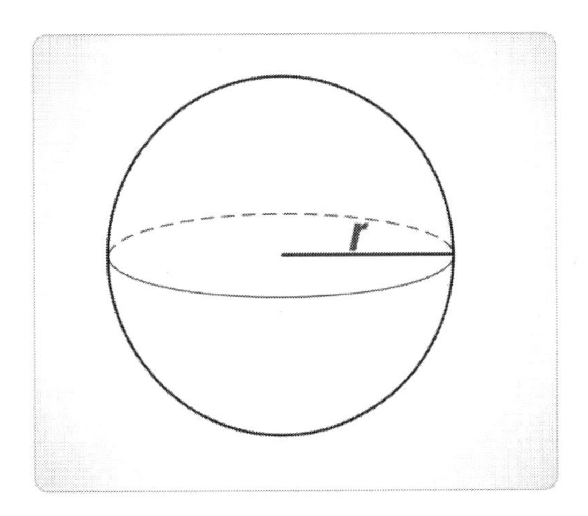

The Pythagorean Theorem

The Pythagorean theorem is an important result in geometry. It states that for right triangles, the sum of the squares of the two shorter sides will be equal to the square of the longest side (also called the *hypotenuse*). The longest side will always be the side opposite to the 90° angle. If this side is called c, and the other two sides are a and b, then the Pythagorean theorem states that $c^2 = a^2 + b^2$. Since lengths are always positive, this also can be written as $c = \sqrt{a^2 + b^2}$.

A diagram to show the parts of a triangle using the Pythagorean theorem is below.

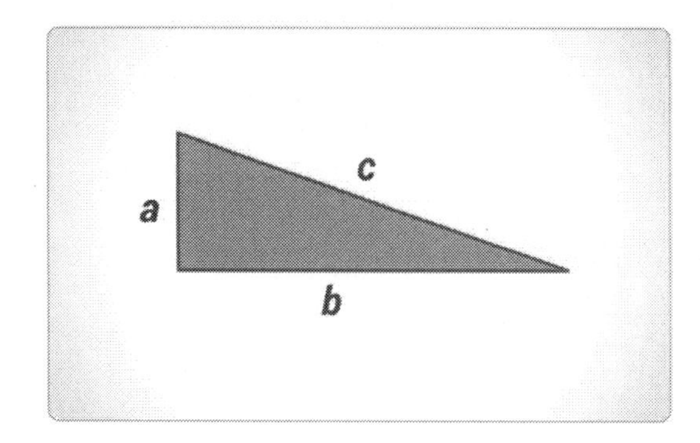

As an example of the theorem, suppose that Shirley has a rectangular field that is 5 feet wide and 12 feet long, and she wants to split it in half using a fence that goes from one corner to the opposite corner. How long will this fence need to be? To figure this out, note that this makes the field into two right triangles, whose hypotenuse will be the fence dividing it in half. Therefore, the fence length will be given by $\sqrt{5^2 + 12^2} = \sqrt{169} = 13$ feet long.

Graphs of Algebraic Functions

A graph can shift in many ways. To shift it horizontally, a constant can be added to all the x variables. Replacing x with $(x + a)$ will shift the graph to the left by a. If a is negative, this shifts the graph to the right. Similarly, vertical shifts occur by adding a constant to each of the y variables. Replacing y by $(y + a)$ will shift the graph up by a. If a is negative, then it shifts the graph down.

A graph can also stretch and shrink the graph in the horizontal and vertical directions. To stretch by a (positive) factor of k horizontally, all instances of x are replaced with $\frac{x}{k}$. To stretch vertically by k, all instances of y are replaced with $\frac{y}{k}$.

The graph can be reflected over the y-axis by replacing all instances of x with $(-x)$. The graph can also be reflected over the x-axis by replacing all instances of y with $(-y)$.

Applications and Other Algebra Topics

Complex Numbers

Some types of equations can be solved to find real answers, but this is not the case for all equations. For example, $x^2 = k$ can be solved when k is non-negative, but it has no real solutions when k is negative. Equations do have solutions if complex numbers are allowed.

Complex numbers are defined in the following manner: every complex number can be written as $a + bi$, where $i^2 = -1$. Thus, the solutions to the equation $x^2 = -1$ are $\pm i$.

In order to find roots of negative numbers more generally, the properties of roots (or of exponents) are used. For example, $\sqrt{-4} = \sqrt{-1}\sqrt{4} = \pm 2i$. All arithmetic operations can be performed with complex numbers, where i is like any other constant. The value of i^2 can replace -1.

Series and Sequences

A *sequence* is an infinite, ordered list of numbers and a function whose domain is an infinite subset of non-negative integers. Usually, the domain will be all non-negative integers, or else all positive integers, but sometimes, it is convenient to use a smaller subset for some sequences. Although it would be possible to use function notation, it is customary to write sequences a little bit differently.

Given a sequence defined by some function, the value of this function is written on *n* as a_n and denoted in the entire sequence as $\{a_n\}$. The sequence can be written $a_1, a_2, a_3, \ldots a_n, \ldots$, keeping in mind this list continues infinitely. The a_n values are called the *terms* of the sequence.

In some cases, a formula for a_n is an expression that only involves *n* and constants. In some other cases, an expression for a_n involves only *n*, constants, and a_{n-1}. This latter case is called a *recursive* definition for the sequences.

An *infinite sum* or simply *sum* of a sequence $\{a_n\}$ is a sequence $\{s_n\}$, where $s_n = a_1 + a_2 + \cdots + a_n$. Some sequences have the property of getting closer to one particular value. The value is the *limit* of the sequence.

If the limit of a sequence $\{a_n\}$ is *L*, this means that for any positive real number δ, there is a value of *M*, as long as $n > M$, $|a_n - L| < \delta$. This is just a very formal way of saying that for any real number positive (real number δ), there is a point where all the remaining values are within δ of *L* in the sequence. This just means, on the whole, getting closer to *L* as the sequence ends. This is a *limit* of a sequence $\lim_{n\to\infty} a_n$.

If a sequence has a limit, the sequence *converges*. On the other hand, if the absolute value $|a_n|$ keeps on getting bigger, the sequence *diverges.*

Some sequences do not converge or diverge. For example, the sequence $a_n = \begin{cases} 1, n \text{ odd} \\ -1, n \text{ even} \end{cases}$, flips back and forth between 1 and -1. This sequence never converges, since it keeps bouncing back and forth. However, it does not diverge, either, since the absolute value is never bigger than 1.

In order to find the limit of a sequence, we can use the following rules:

- $\lim_{n\to\infty} k = k$ for all real numbers *k*.
- $\lim_{n\to\infty} \frac{1}{n} = 0.$
- $\lim_{n\to\infty} n = \infty.$
- $\lim_{n\to\infty} \frac{k}{n^p} = 0$ when *k* is real and *p* is a positive rational number.
- $\lim_{n\to\infty} (a_n + b_n) = \lim_{n\to\infty} a_n + \lim_{n\to\infty} b_n$
- $\lim_{n\to\infty} (a_n - b_n) = \lim_{n\to\infty} a_n - \lim_{n\to\infty} b_n$

- $\lim_{n\to\infty}(a_n \cdot b_n) = \lim_{n\to\infty} a_n \cdot \lim_{n\to\infty} b_n$. As a special case, $\lim_{n\to\infty} ka_n = k \lim_{n\to\infty} a_n$
- $\lim_{n\to\infty}\left(\frac{a_n}{b_n}\right) = \frac{\lim_{n\to\infty} a_n}{\lim_{n\to\infty} b_n}$ when $\lim_{n\to\infty} b_n$ is not 0.

A sequence is called *monotonic* if the terms of the sequence never decrease or if they never increase. In other words, $\{a_n\}$ is monotonic if one of two things happen: either $a_n \geq a_m$ whenever $n > m$ (in this case, the sequences is also called *non-decreasing*), or else $a_n \leq a_m$ whenever $n > m$ (in this case, the sequence is also called *non-increasing*).

A sequence is said to be *bounded above* by *k* if, for any value of *n*, every $a_n \leq k$, and *bounded below* by *k* if $a_n \geq k$. Every non-decreasing sequence that is bounded above converges to some real number. Every non-increasing sequence that is bounded below converges to some real number.

An *arithmetic sequence* is a sequence where the next term is obtained from the previous term by adding a specific quantity, *k*. In other words, $a_{n+1} = a_n + k$. Another way of writing this out is the sequence $a_1, a_1 + k, a_1 + 2k, \dots, a_1 + (n-1)k$ .... That is, $a_n = a_1 + (n-1)k$. The *sum* of the first *n* terms of an arithmetic sequence is $s_n = \frac{n}{2}(a_1 + a_n)$.

A *geometric sequence* (or a *geometric progression*) is a sequence in which for some specific quantity *r*, $a_{n+1} = ra_n$. Another way of writing this is that the sequence is $a_1, a_1r, a_1r^2, \dots, a_1r^{n-1}$ .... The general formula is $a_n = a_1r^{n-1}$. The sum of the first *n* terms of a geometric sequence is $s_n = \frac{a_1(1-r^n)}{1-r}$.

A *series* or *infinite series* is a sequence $\{s_n\}$, whose *n*-th term is the sum of the first *n* terms of some sequence $\{a_n\}$. Thus, $s_n = a_1 + a_2 + \cdots + a_n$ can also be written as $\sum_{m=1}^{n} a_m$.

Each s_n is the sum of the first n terms and is called the *n-th partial sum.* The *infinite sum* (or simply, *sum*) of a sequence $\{a_n\}$ is the limit of the sequence $\{s_n\}$, also written as $\sum_{n=1}^{\infty} a_n$.

A series can converge, diverge, or neither, just like a sequence. The rules for finding whether or not a series converges or diverges are identical to the rules for sequences. When a series is being added or subtracted, or multiplied by constants, it obeys the following rules:

$$\sum_{n=1}^{\infty}(a_n + b_n) = \sum_{n=1}^{\infty} a_n + \sum_{n=1}^{\infty} b_n$$

$$\sum_{n=1}^{\infty}(a_n - b_n) = \sum_{n=1}^{\infty} a_n - \sum_{n=1}^{\infty} b_n$$

$$\sum_{n=1}^{\infty} ka_n = k\sum_{n=1}^{\infty} a_n$$

A *geometric series* is a sum of a geometric sequence: $\sum_{n=1}^{\infty} ar^{n-1} = a_1 + a_2r + \cdots + a_nr^{n-1} + \cdots$. In such a series, when $|r| \geq 1$, the series diverges. However, when $|r| < 1$, then $\sum_{n=1}^{\infty} ar^{n-1} = \frac{a}{1-r}$.

It's important to note that whenever a sum $\sum_{n=1}^{\infty} a_n$ converges, the sequence $\{a_n\}$ has a limit of 0: $\lim_{n\to\infty} a_n = 0$. This is one possible test to see whether or not a series converges. However, just because this limit is zero does not mean that the sum diverges, so the test only works in one direction.

Determinants

A *matrix* is a rectangular arrangement of numbers in rows and columns. The *determinant* of a matrix is a special value that can be calculated for any square matrix.

Using the *square 2 x 2 matrix* $\begin{bmatrix} a & b \\ c & d \end{bmatrix}$, the determinant is $ad - bc$.

For example, the determinant of the matrix $\begin{bmatrix} -5 & 1 \\ 3 & 4 \end{bmatrix}$ is *-5(4) – 1(3) = -20 – 3 = -23.*

Using a *3 x 3 matrix* $\begin{bmatrix} a & b & c \\ d & e & f \\ g & h & i \end{bmatrix}$, the determinant is $a(ei - fh) - b(di - fg) + c(dh - eg)$.

For example, the determinant of the matrix $\begin{bmatrix} 2 & 0 & 1 \\ -1 & 3 & 2 \\ 2 & -2 & -1 \end{bmatrix}$ is

$$2\big(3(-1) - 2(-2)\big) - 0\big(-1(-1) - 2(2)\big) + 1\big(-1(-2) - 3(2)\big)$$

$$= 2(-3 + 4) - 0(1 - 4) + 1(2 - 6)$$

$$= 2(1) - 0(-3) + 1(-4)$$

$$= 2 - 0 - 4 = -2$$

The pattern continues for larger square matrices.

Permutations and Combinations

The *factorial* is defined for non-negative integers. The factorial of *n* is written as $n!$ For positive integers, the factorial is defined as the product of all positive integers up to *n*. So, $n! = 1 \cdot 2 \cdot \ldots \cdot n$.

For zero, $0! = 1$. The reason for zero being a special case is two-fold. First, the relation is always $n! = n \cdot (n-1)!$ when the right hand side is defined. Second, it makes the choice functions below work out correctly.

The *combinatorial choice function* indicates how many distinct ways one can choose to pick out *k* objects from a set of *n* objects. The choice function is written as $\binom{n}{k}$—read as "*n* choose *k*"—and is given by $\binom{n}{k} = \frac{n!}{k! \cdot (n-k)!}$.

As an example, suppose a person wanted to choose three shirts out of five shirts to take on a trip. How many ways can this be done?

The answer is given by computing 5 choose 3. $\binom{5}{3} = \frac{5!}{3! \cdot (5-3)!} = \frac{5 \cdot 4 \cdot 3 \cdot 2 \cdot 1}{3 \cdot 2 \cdot 1 \cdot 2 \cdot 1}$. At this stage, all the indicated multiplications can be calculated, but a number of common factors cancel first. (The appearance of many common factors is normal when computing choice functions.)

The expression simplifies to, $\frac{5 \cdot 4 \cdot 3 \cdot 2 \cdot 1}{3 \cdot 2 \cdot 1 \cdot 2 \cdot 1} = \frac{5 \cdot 4}{2 \cdot 1} = \frac{20}{2} = 10$.

There are ten different combinations of three shirts.

Using the choice function, it can be calculated how many ways to order a set of objects. Ordering a set of *n* objects requires choosing a first object out of *n* objects, then picking a second object out of the remaining *n* – 1 objects. Then a third object out of the *n* – 2 objects can be picked, and so on. Therefore, the total number of ways to order *n* objects will be $\binom{n}{1}\binom{n-1}{1} \dots \binom{1}{1}$

Fractions and Word Problems

Work word problems are examples of people working together in a situation that uses fractions.

Example

One painter can paint a designated room in 6 hours, and a second painter can paint the same room in 5 hours. How long will it take them to paint the room if they work together?

The first painter paints $\frac{1}{6}$of the room in an hour, and the second painter paints $\frac{1}{5}$of the room in an hour.

Together, they can paint $\frac{1}{x}$ of the room in an hour. The equation is the sum of the painters rate equal to the total job or$\frac{1}{6} + \frac{1}{5} = \frac{1}{x}$.

The equation can be solved by multiplying all terms by a common denominator of 30*x* with a result of $5x + 6x = 30$.

The left side can be added together to get 11*x*, and then divide by 11 for a solution of $\frac{30}{11}$or about 2.73 hours.

Translating Words into Math

When asked rewrite a mathematical expression as a situation or translate a word problem into an expression, look for a series of key words indicating addition, subtraction, multiplication, or division:

> **Addition**: add, altogether, together, plus, increased by, more than, in all, sum, and total
>
> **Subtraction**: minus, less than, difference, decreased by, fewer than, remain, and take away
>
> **Multiplication**: *times*, *twice*, *of*, *double*, and *triple*
>
> **Division**: divided by, cut up, half, quotient of, split, and shared equally

If a question asks to give words to a mathematical expression and says "equals," then an = sign must be included in the answer. Similarly, "less than or equal to" is expressed by the inequality symbol ≤, and "greater than or equal" to is expressed as ≥. Furthermore, "less than" is represented by <, and "greater than" is expressed by >.

Solving Real Word One- or Multi-Step Problems with Rational Numbers

Word problems can appear daunting, but don't let the verbiage psych you out. No matter the scenario or specifics, the key to answering them is to translate the words into a math problem. Always keep in mind what the question is asking and what operations could lead to that answer. The following word problems highlight the most commonly tested question types.

Working with Money

Walter's Coffee Shop sells a variety of drinks and breakfast treats.

Price List	
Hot Coffee	$2.00
Slow-Drip Iced Coffee	$3.00
Latte	$4.00
Muffin	$2.00
Crepe	$4.00
Egg Sandwich	$5.00

Costs	
Hot Coffee	$0.25
Slow-Drip Iced Coffee	$0.75
Latte	$1.00
Muffin	$1.00
Crepe	$2.00
Egg Sandwich	$3.00

Walter's utilities, rent, and labor costs him $500 per day. Today, Walter sold 200 hot coffees, 100 slow-drip iced coffees, 50 lattes, 75 muffins, 45 crepes, and 60 egg sandwiches. What was Walter's total profit today?

To accurately answer this type of question, determine the total cost of making his drinks and treats, then determine how much revenue he earned from selling those products. After arriving at these two totals, the profit is measured by deducting the total cost from the total revenue.

Walter's costs for today:

Item	Quantity	Cost Per Unit	Total Cost
Hot Coffee	200	$0.25	$50
Slow-Drip Iced Coffee	100	$0.75	$75
Latte	50	$1.00	$50
Muffin	75	$1.00	$75
Crepe	45	$2.00	$90
Egg Sandwich	60	$3.00	$180
Utilities, rent, and labor			$500
Total Costs			$1,020

Walter's revenue for today:

Item	Quantity	Revenue Per Unit	Total Revenue
Hot Coffee	200	\$2.00	\$400
Slow-Drip Iced Coffee	100	\$3.00	\$300
Latte	50	\$4.00	\$200
Muffin	75	\$2.00	\$150
Crepe	45	\$4.00	\$180
Egg Sandwich	60	\$5.00	\$300
Total Revenue			\$1,530

Walter's Profit = *Revenue* – *Costs* = \$1,530 – \$1,020 = \$510

This strategy is applicable to other question types. For example, calculating salary after deductions, balancing a checkbook, and calculating a dinner bill are common word problems similar to business planning. Just remember to use the correct operations. When a balance is increased, use addition. When a balance is decreased, use subtraction. Common sense and organization are your greatest assets when answering word problems.

Story Problems

Story problems describe scenarios where algebra is needed to solve real life problems.

It's important to understand what is being asked and to properly set up the initial equation. Always read the entire problem through, and then separate out what information is given in the statement. Decide what you are being asked to find and label each quantity with a variable or constant. Then write an equation to determine the unknown variable. Remember to label answers; sometimes knowing what the answers' units will be can help eliminate other possible solutions.

<u>Example</u>
A store is having a spring sale, where everything is 70% off. You have \$45.00 to spend. A jacket is regularly priced at \$80.00. Do you have enough to buy the jacket and a pair of gloves, regularly priced at \$20.00?

There are two ways to approach this.

Method 1:

Set up the equations to find the sale prices: the original price minus the amount discounted.
$80.00 - ($80.00 (0.70)) = sale cost of the jacket.
$20.00 – ($20.00 (0.70)) = sale cost of the gloves.
Solve for the sale cost.
$24.00 = sale cost of the jacket.
$6.00 = sale cost of the gloves.
Determine if you have enough money for both.
$24.00 + $6.00 = total sale cost.
$30.00 is less than $45.00, so you can afford to purchase both.

Method 2:

Determine the percent of the original price that you will pay.
100% – 70% = 30%
Set up the equations to find the sale prices.
$80.00 (0.30) = cost of the jacket.
$20.00 (0.30) = cost of the gloves.
Solve.
$24.00 = cost of the jacket.
$6.00 = cost of the gloves.
Determine if you have enough money for both.
$24.00 + $6.00 = total sale cost.
$30.00 is less than $45.00, so you can afford to purchase both.

Example

Mary and Dottie team up to mow neighborhood lawns. If Mary mows 2 lawns per hour and the two of them can mow 17.5 lawns in 5 hours, how many lawns does Dottie mow per hour?

Given rate for Mary.

$$Mary = \frac{2\ lawns}{1\ hour}$$

Unknown rate of D for Dottie.

$$Dottie = \frac{D\ lawns}{1\ hour}$$

Given rate for both.

$$Total\ mowed\ together = \frac{17.5\ lawns}{5\ hours}$$

Set up the equation for what is being asked.

$$Mary + Dottie = total\ together.$$

Fill in the givens.

$$2 + D = \frac{17.5}{5}$$

Divide.

$$2 + D = 3.5$$

Subtract 2 from both sides to isolate the variable.

$$2 - 2 + D = 3.5 - 2$$

Solve and label Dottie's mowing rate.

$$D = 1.5\ lawns\ per\ hour$$

Solving Real-World Problems Involving Percentages

Questions dealing with percentages can be difficult when they are phrased as word problems. These word problems almost always come in three varieties. The first type will ask to find what percentage of some number will equal another number. The second asks to determine what number is some percentage of another given number. The third will ask what number another number is a given percentage of.

One of the most important parts of correctly answering percentage word problems is to identify the numerator and the denominator. This fraction can then be converted into a percentage, as described above.

The following word problem shows how to make this conversion:

A department store carries several different types of footwear. The store is currently selling 8 athletic shoes, 7 dress shoes, and 5 sandals. What percentage of the store's footwear are sandals?

First, calculate what serves as the 'whole', as this will be the denominator. How many total pieces of footwear does the store sell? The store sells 20 different types (8 athletic + 7 dress + 5 sandals).

Second, what footwear type is the question specifically asking about? Sandals. Thus, 5 is the numerator.

Third, the resultant fraction must be expressed as a percentage. The first two steps indicate that $\frac{5}{20}$ of the footwear pieces are sandals. This fraction must now be converted into a percentage:

$$\frac{5}{20} \times \frac{5}{5} = \frac{25}{100} = 25\%$$

Solving Real-World Problems Involving Proportions

Much like a scale factor can be written using an equation like $2A = B$, a *relationship* is represented by the equation $Y = kX$. X and Y are proportional because as values of X increase, the values of Y also increase. A relationship that is inversely proportional can be represented by the equation $Y = \frac{k}{X}$, where the value of Y decreases as the value of x increases and vice versa.

Proportional reasoning can be used to solve problems involving ratios, percentages, and averages. Ratios can be used in setting up proportions and solving them to find unknowns. For example, if student completes an average of 10 pages of math homework in 3 nights, how long would it take the student to complete 22 pages? Both ratios can be written as fractions. The second ratio would contain the unknown.

The following proportion represents this problem, where x is the unknown number of nights:

$$\frac{10\ pages}{3\ nights} = \frac{22\ pages}{x\ nights}$$

Solving this proportion entails cross-multiplying and results in the following equation: $10x = 22 * 3$. Simplifying and solving for x results in the exact solution: $x = 6.6\ nights$. The result would be rounded up to 7 because the homework would be actually be completed on the 7th night.

The following problem uses ratios involving percentages:

If 20% of the class is girls and 30 students are in the class, how many girls are in the class?

To set up this problem, it is helpful to use the common proportion: $\frac{\%}{100} = \frac{is}{of}$. Within the proportion, % is the percentage of girls, 100 is the total percentage of the class, *is* is the number of girls, and *of* is the total number of students in the class. Most percentage problems can be written using this language. To solve this problem, the proportion should be set up as $\frac{20}{100} = \frac{x}{30}$, and then solved for x. Cross-multiplying results in the equation $20 * 30 = 100x$, which results in the solution $x = 6$. There are 6 girls in the class.

Problems involving volume, length, and other units can also be solved using ratios. For example, a problem may ask for the volume of a cone to be found that has a radius, $r = 7m$ and a height, $h = 16m$. Referring to the formulas provided on the test, the volume of a cone is given as: $V = \pi r^2 \frac{h}{3}$, where r is the radius, and h is the height. Plugging $r = 7$ and $h = 16$ into the formula, the following is obtained: $V = \pi(7^2)\frac{16}{3}$. Therefore, volume of the cone is found to be approximately 821m^3. Sometimes, answers in different units are sought. If this problem wanted the answer in liters, 821m^3 would need to be converted. Using the equivalence statement 1m^3 = 1000L, the following ratio would be used to solve for liters: $821\text{m}^3 * \frac{1000L}{1m^3}$. Cubic meters in the numerator and denominator cancel each other out, and the answer is converted to 821,000 liters, or $8.21 * 10^5$ L.

Other conversions can also be made between different given and final units. If the temperature in a pool is 30°C, what is the temperature of the pool in degrees Fahrenheit? To convert these units, an equation is used relating Celsius to Fahrenheit. The following equation is used: $T_{°F} = 1.8T_{°C} + 32$. Plugging in the given temperature and solving the equation for T yields the result: $T_{°F} = 1.8(30) + 32 = 86°F$. Both units in the metric system and U.S. customary system are widely used.

Here are some more examples of how to solve for proportions:

1) $\frac{75\%}{90\%} = \frac{25\%}{x}$

To solve for x, the fractions must be cross multiplied: ($75\%x = 90\% \times 25\%$). To make things easier, let's convert the percentages to decimals: ($0.9 \times 0.25 = 0.225 = 0.75x$). To get rid of x's co-efficient, each side must be divided by that same coefficient to get the answer $x = 0.3$. The question could ask for the answer as a percentage or fraction in lowest terms, which are 30% and $\frac{3}{10}$, respectively.

2) $\frac{x}{12} = \frac{30}{96}$

Cross-multiply: $96x = 30 \times 12$

Multiply: $96x = 360$

Divide: $x = 360 \div 96$

Answer: $x = 3.75$

3) $\frac{0.5}{3} = \frac{x}{6}$

Cross-multiply: $3x = 0.5 \times 6$

Multiply: $3x = 3$

Divide: $x = 3 \div 3$

Answer: $x = 1$

You may have noticed there's a faster way to arrive at the answer. If there is an obvious operation being performed on the proportion, the same operation can be used on the other side of the proportion to solve for x. For example, in the first practice problem, 75% became 25% when divided by 3, and upon doing the same to 90%, the correct answer of 30% would have been found with much less legwork. However, these questions aren't always so intuitive, so it's a good idea to work through the steps, even if the answer seems apparent from the outset.

Solving Real-World Problems Involving Ratios and Rates of Change

Ratios are used to show the relationship between two quantities. The ratio of oranges to apples in the grocery store may be 3 to 2. That means that for every 3 oranges, there are 2 apples. This comparison can be expanded to represent the actual number of oranges and apples. Another example may be the number of boys to girls in a math class. If the ration of boys to girls is given as 2 to 5, that means there are 2 boys to every 5 girls in the class. Ratios can also be compared if the units in each ratio are the same. The ratio of boys to girls in the math class can be compared to the ratio of boys to girls in a science class by stating which ratio is higher and which is lower.

Rates are used to compare two quantities with different units. *Unit rates* are the simplest form of rate. With unit rates, the denominator in the comparison of two units is one. For example, if someone can type at a rate of 1000 words in 5 minutes, then his or her unit rate for typing is $\frac{1000}{5} = 200$ words in one minute or 200 words per minute. Any rate can be converted into a unit rate by dividing to make the denominator one. 1000 words in 5 minutes has been converted into the unit rate of 200 words per minute.

Ratios and rates can be used together to convert rates into different units. For example, if someone is driving 50 kilometers per hour, that rate can be converted into miles per hour by using a ratio known as the *conversion factor*. Since the given value contains kilometers and the final answer needs to be in miles, the ratio relating miles to kilometers needs to be used. There are 0.62 miles in 1 kilometer. This, written as a ratio and in fraction form, is $\frac{0.62\ miles}{1\ km}$. To convert 50km/hour into miles per hour, the following conversion needs to be set up: $\frac{50\ km}{hour} * \frac{0.62\ miles}{1\ km} = 31\ miles\ per\ hour.$

The ratio between two similar geometric figures is called the *scale factor*. For example, a problem may depict two similar triangles, A and B. The scale factor from the smaller triangle A to the larger triangle B is given as 2 because the length of the corresponding side of the larger triangle, 16, is twice the corresponding side on the smaller triangle, 8. This scale factor can also be used to find the value of a missing side, x, in triangle A. Since the scale factor from the smaller triangle (A) to larger one (B) is 2, the larger corresponding side in triangle B (given as 25), can be divided by 2 to find the missing side in A (x = 12.5). The scale factor can also be represented in the equation $2A = B$ because two times the lengths of A gives the corresponding lengths of B. This is the idea behind similar triangles.

Unit Rate

Unit rate word problems will ask to calculate the rate or quantity of something in a different value. For example, a problem might say that a car drove a certain number of miles in a certain number of minutes and then ask how many miles per hour the car was traveling. These questions involve solving proportions. Consider the following examples:

1) Alexandra made $96 during the first 3 hours of her shift as a temporary worker at a law office. She will continue to earn money at this rate until she finishes in 5 more hours. How much does Alexandra make per hour? How much will Alexandra have made at the end of the day?

This problem can be solved in two ways. The first is to set up a proportion, as the rate of pay is constant. The second is to determine her hourly rate, multiply the 5 hours by that rate, and then add the $96.

To set up a proportion, put the money already earned over the hours already worked on one side of an equation. The other side has x over 8 hours (the total hours worked in the day). It looks like this: $\frac{96}{3} = \frac{x}{8}$. Now, cross-multiply to get $768 = 3x$. To get x, divide by 3, which leaves $x = 256$. Alternatively, as x is the numerator of one of the proportions, multiplying by its denominator will reduce the solution by one step. Thus, Alexandra will make $256 at the end of the day. To calculate her hourly rate, divide the total by 8, giving $32 per hour.

Alternatively, it is possible to figure out the hourly rate by dividing $96 by 3 hours to get $32 per hour. Now her total pay can be figured by multiplying $32 per hour by 8 hours, which comes out to $256.

2) Jonathan is reading a novel. So far, he has read 215 of the 335 total pages. It takes Jonathan 25 minutes to read 10 pages, and the rate is constant. How long does it take Jonathan to read one page? How much longer will it take him to finish the novel? Express the answer in time.

To calculate how long it takes Jonathan to read one page, divide the 25 minutes by 10 pages to determine the page per minute rate. Thus, it takes 2.5 minutes to read one page.

Jonathan must read 120 more pages to complete the novel. (This is calculated by subtracting the pages already read from the total.) Now, multiply his rate per page by the number of pages. Thus, $120 \div 2.5 = 300$. Expressed in time, 300 minutes is equal to 5 hours.

3) At a hotel, $\frac{4}{5}$ of the 120 rooms are booked for Saturday. On Sunday, $\frac{3}{4}$ of the rooms are booked. On which day are more of the rooms booked, and by how many more?

The first step is to calculate the number of rooms booked for each day. Do this by multiplying the fraction of the rooms booked by the total number of rooms.

Saturday: $\frac{4}{5} \times 120 = \frac{4}{5} \times \frac{120}{1} = \frac{480}{5} = 96$ rooms

Sunday: $\frac{3}{4} \times 120 = \frac{3}{4} \times \frac{120}{1} = \frac{360}{4} = 90$ rooms

Thus, more rooms were booked on Saturday by 6 rooms.

4) In a veterinary hospital, the veterinarian-to-pet ratio is 1:9. The ratio is always constant. If there are 45 pets in the hospital, how many veterinarians are currently in the veterinary hospital?

Set up a proportion to solve for the number of veterinarians: $\frac{1}{9} = \frac{x}{45}$

Cross-multiplying results in $9x = 45$, which works out to 5 veterinarians.

Alternatively, as there are always 9 times as many pets as veterinarians, is it possible to divide the number of pets (45) by 9. This also arrives at the correct answer of 5 veterinarians.

5) At a general practice law firm, 30% of the lawyers work solely on tort cases. If 9 lawyers work solely on tort cases, how many lawyers work at the firm?

First, solve for the total number of lawyers working at the firm, which will be represented here with x. The problem states that 9 lawyers work solely on torts cases, and they make up 30% of the total lawyers at the firm. Thus, 30% multiplied by the total, x, will equal 9. Written as equation, this is: $30\% \times x = 9$.

It's easier to deal with the equation after converting the percentage to a decimal, leaving $0.3x = 9$. Thus, $x = \frac{9}{0.3} = 30$ lawyers working at the firm.

6) Xavier was hospitalized with pneumonia. He was originally given 35mg of antibiotics. Later, after his condition continued to worsen, Xavier's dosage was increased to 60mg. What was the percent increase of the antibiotics? Round the percentage to the nearest tenth.

An increase or decrease in percentage can be calculated by dividing the difference in amounts by the original amount and multiplying by 100. Written as an equation, the formula is:

$$\frac{new\ quantity - old\ quantity}{old\ quantity} \times 100$$

Here, the question states that the dosage was increased from 35mg to 60mg, so these are plugged into the formula to find the percentage increase.

$$\frac{60 - 35}{35} \times 100 = \frac{25}{35} \times 100 = .7142 \times 100 = 71.4\%$$

Functions and Trigonometry

Relations and Functions

First, it's important to understand the definition of a *relation*. Given two variables, *x* and *y*, which stand for unknown numbers, a *relation* between *x* and *y* is an object that splits all of the pairs (*x*, *y*) into those for which the relation is true and those for which it is false. For example, consider the relation of $x^2 = y^2$. This relationship is true for the pair (1, 1) and for the pair (-2, 2), but false for (2, 3). Another example of a relation is $x \leq y$. This is true whenever *x* is less than or equal to *y*.

A *function* is a special kind of relation where, for each value of *x*, there is only a single value of *y* that satisfies the relation. So, $x^2 = y^2$ is *not* a function because in this case, if *x* is 1, *y* can be either 1 or -1: the pair (1, 1) and (1, -1) both satisfy the relation. More generally, for this relation, any pair of the form $(a, \pm a)$ will satisfy it. On the other hand, consider the following relation: $y = x^2 + 1$. This is a function because for each value of *x*, there is a unique value of *y* that satisfies the relation. Notice, however, there are multiple values of *x* that give us the same value of *y*. This is perfectly acceptable for a function. Therefore, *y* is a function of *x*.

To determine if a relation is a function, check to see if every *x* value has a unique corresponding *y* value.

A function can be viewed as an object that has *x* as its input and outputs a unique *y*-value. It is sometimes convenient to express this using *function notation*, where the function itself is given a name, often *f*. To emphasize that *f* takes *x* as its input, the function is written as $f(x)$. In the above example, the equation could be rewritten as $f(x) = x^2 + 1$. To write the value that a function yields for some specific value of *x*, that value is put in place of *x* in the function notation. For example, $f(3)$ means the value that the function outputs when the input value is 3. If $f(x) = x^2 + 1$, then $f(3) = 3^2 + 1 = 10$.

A function can also be viewed as a table of pairs (*x*, *y*), which lists the value for *y* for each possible value of *x*.

The set of all possible values for *x* in $f(x)$ is called the *domain* of the function, and the set of all possible outputs is called the *range* of the function. Note that usually the domain is assumed to be all real numbers, except those for which the expression for $f(x)$ is not defined, unless the problem specifies otherwise. An example of how a function might not be defined is in the case of $f(x) = \frac{1}{x+1}$, which is not defined when $x = -1$ (which would require dividing by zero). Therefore, in this case the domain would be all real numbers except $x = -1$.

If *y* is a function of *x*, then *x* is the *independent variable* and *y* is the *dependent variable*. This is because in many cases, the problem will start with some value of *x* and then see how *y* changes depending on this starting value.

Evaluating Functions

To evaluate functions, plug in the given value everywhere the variable appears in the expression for the function. For example, find $g(-2)$ where $g(x) = 2x^2 - \frac{4}{x}$. To complete the problem, plug in -2 in the following way: $g(-2) = 2(-2)^2 - \frac{4}{-2} = 2 \cdot 4 + 2 = 8 + 2 = 10$.

Graphing Functions and Relations

To graph relations and functions, the Cartesian plane is used. This means to think of the plane as being given a grid of squares, with one direction being the x-axis and the other direction the y-axis. Generally, the independent variable is placed along the horizontal axis, and the dependent variable is placed along the vertical axis. Any point on the plane can be specified by saying how far to go along the x-axis and how far along the y-axis with a pair of numbers (x, y). Specific values for these pairs can be given names such as $C = (-1, 3)$. Negative values mean to move left or down; positive values mean to move right or up. The point where the axes cross one another is called the *origin*. The origin has coordinates $(0, 0)$ and is usually called *O* when given a specific label. An illustration of the Cartesian plane, along with graphs of $(2, 1)$ and $(-1, -1)$, are below.

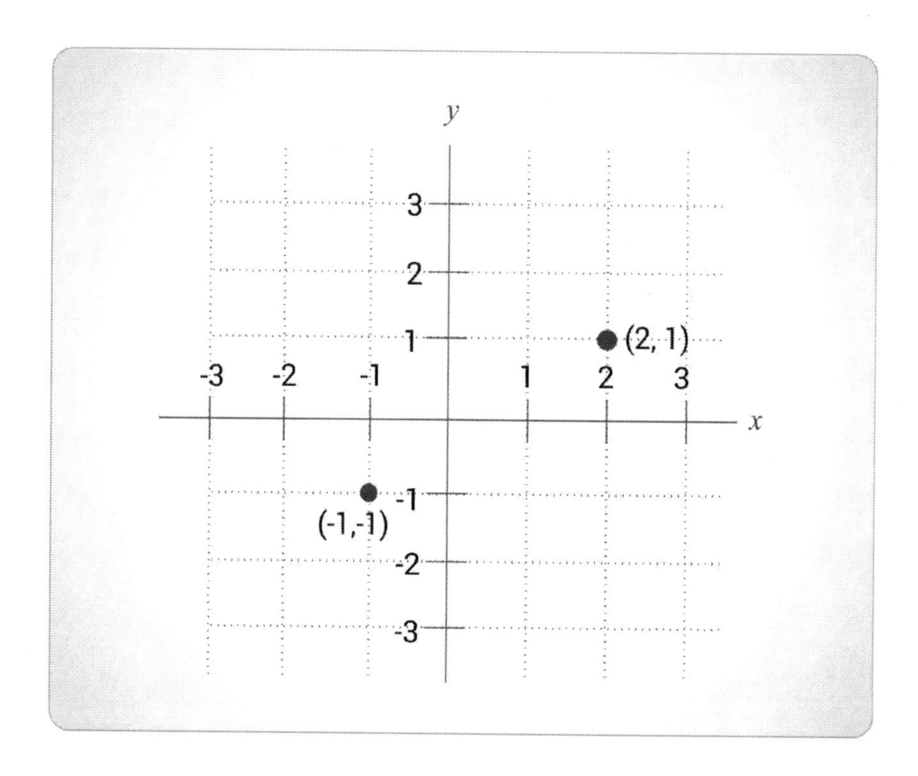

Relations also can be graphed by marking each point whose coordinates satisfy the relation. If the relation is a function, then there is only one value of y for any given value of x. This leads to the *vertical line test*: if a relation is graphed, then the relation is a function if every vertical line touches the graph at either zero or one point. Conversely, when graphing a function, then every vertical line will touch the graph at no points or just one point.

Rate of Change

The rate of change for a linear function is constant and can be determined based on a few representations. One method is to place the equation in slope-intercept form: $y = mx + b$. Thus, m is the slope, and b is the y-intercept. In the graph below, the equation is $y = x + 1$, where the slope is 1 and the y-intercept is 1. For every vertical change of 1 unit, there is a horizontal change of 1 unit.

The x-intercept is -1, which is the point where the line crosses the x-axis:

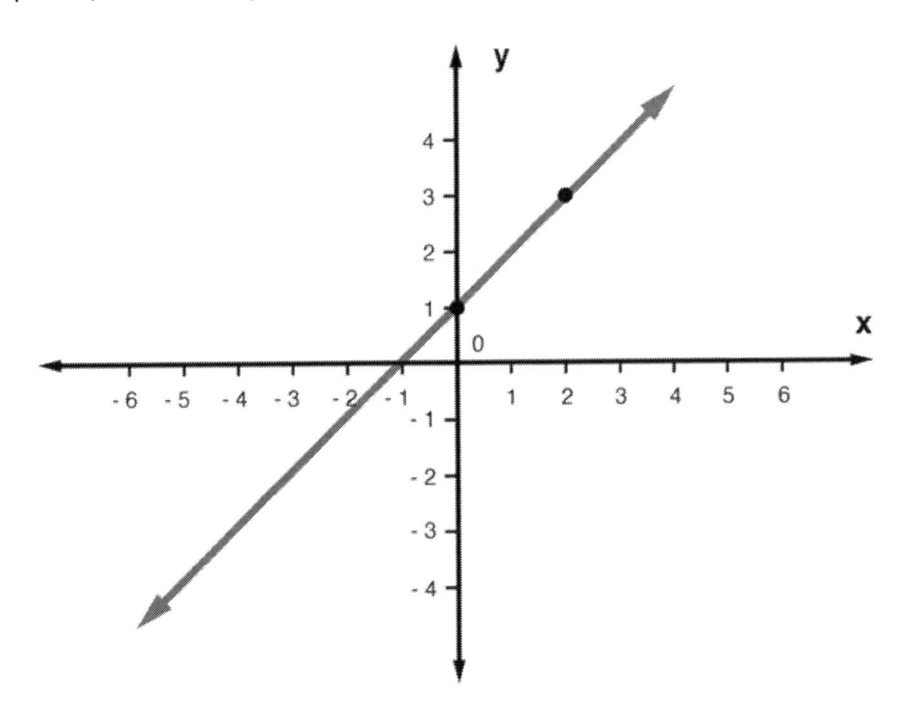

Algebraic Functions

A function is called *algebraic* if it is built up from polynomials by adding, subtracting, multplying, dividing, and taking radicals. This means that, for example, the variable can never appear in an exponent. Thus, polynomials and rational functions are algebraic, but exponential functions are not algebraic. It turns out that logarithms and trigonometric functions are not algebraic either.

A function of the form $f(x) = a_n x^n + a_{n-1} x^{n-1} + a_{n-2} x^{n-2} + \cdots + a_1 x + a_0$ is called a *polynomial function*. The value of *n* is called the *degree* of the polynomial. In the case where $n = 1$, it is called a *linear function*. In the case where $n = 2$, it is called a *quadratic function*. In the case where $n = 3$, it is called a *cubic function*.

When *n* is even, the polynomial is called *even*, and not all real numbers will be in its range. When *n* is odd, the polynomial is called *odd*, and the range includes all real numbers.

The graph of a quadratic function $f(x) = ax^2 + bx + c$ will be a parabola. To see whether or not the parabola opens up or down, it's necessary to check the coefficient of x^2, which is the value a.

If the coefficient is positive, then the parabola opens upward. If the coefficient is negative, then the parabola opens downward.

The quantity $D = b^2 - 4ac$ is called the *discriminant* of the parabola. If the discriminant is positive, then the parabola has two real zeros. If the discriminant is negative, then it has no real zeros.

If the discriminant is zero, then the parabola's highest or lowest point is on the *x*-axis, and it has a single real zero.

The highest or lowest point of the parabola is called the *vertex*. The coordinates of the vertex are given by the point $(-\frac{b}{2a}, -\frac{D}{4a})$. The roots of a quadratic function can be found with the quadratic formula, which is $x = \frac{-b\pm\sqrt{b^2-4ac}}{2a}$.

A *rational function* is a function $f(x) = \frac{p(x)}{q(x)}$, where *p* and *q* are both polynomials. The domain of *f* will be all real numbers except the (real) roots of *q*.

At these roots, the graph of *f* will have a *vertical asymptote*, unless they are also roots of *p*. Here is an example to consider:

$$p(x) = p_n x^n + p_{n-1}x^{n-1} + p_{n-2}x^{n-2} + \cdots + p_1 x + p_0$$

$$q(x) = q_m x^m + q_{m-1}x^{m-1} + q_{m-2}x^{m-2} + \cdots + q_1 x + q_0$$

When the degree of *p* is less than the degree of *q*, there will be a horizontal asymptote of $y = 0$. If *p* and *q* have the same degree, there will be a horizontal asymptote of $y = \frac{p_n}{q_n}$. If the degree of *p* is exactly one greater than the degree of *q*, then *f* will have an oblique asymptote along the line $y = \frac{p_n}{q_{n-1}}x + \frac{p_{n-1}}{q_{n-1}}$.

Building a Linear Function that Models a Linear Relationship Between Two Quantities

Linear relationships between two quantities can be expressed in two ways: function notation or as a linear equation with two variables. The relationship is referred to as linear because its graph is represented by a line. For a relationship to be linear, both variables must be raised to the first power only.

Function/Linear Equation Notation

A relation is a set of input and output values that can be written as ordered pairs. A function is a relation in which each input is paired with exactly one output. The domain of a function consists of all inputs, and the range consists of all outputs. Graphing the ordered pairs of a linear function produces a straight line. An example of a function would be $f(x) = 4x + 4$, read "*f* of *x* is equal to four times *x* plus four." In this example, the input would be *x* and the output would be f(*x*). Ordered pairs would be represented as (*x*, f(*x*)). To find the output for an input value of 3, 3 would be substituted for *x* into the function as follows: $f(3) = 4(3) + 4$, resulting in $f(3) = 16$. Therefore, the ordered pair $(3, f(3)) = (3, 16)$. Note f(*x*) is a function of *x* denoted by *f*. Functions of *x* could be named *g*(*x*), read "*g* of *x*"; *p*(*x*), read "*p* of *x*"; etc.

A linear function could also be written in the form of an equation with two variables. Typically, the variable *x* represents the inputs and the variable *y* represents the outputs. The variable *x* is considered the independent variable and *y* the dependent variable. The above function would be written as $y = 4x + 4$. Ordered pairs are written in the form (*x*, *y*).

Writing Linear Equations in Two Variables

When writing linear equations in two variables, the process depends on the information given. Questions will typically provide the slope of the line and its *y*-intercept, an ordered pair and the slope, or two ordered pairs.

Given the Slope and Y-Intercept
Linear equations are commonly written in slope-intercept form, $y = mx + b$, where *m* represents the slope of the line and *b* represents the *y*-intercept. The slope is the rate of change between the variables, usually expressed as a whole number or fraction. The *y*-intercept is the value of *y* when *x* = 0 (the point where the line intercepts the *y*-axis on a graph). Given the slope and *y*-intercept of a line, the values are substituted for *m* and *b* into the equation. A line with a slope of ½ and *y*-intercept of -2 would have an equation $y = ½x - 2$.

Given an Ordered Pair and the Slope
The point-slope form of a line, $y - y_1 = m(x - x_1)$, is used to write an equation when given an ordered pair (point on the equation's graph) for the function and its rate of change (slope of the line). The values for the slope, *m*, and the point (x_1, y_1) are substituted into the point-slope form to obtain the equation of the line. A line with a slope of 3 and an ordered pair (4, -2) would have an equation $y - (-2) = 3(x - 4)$. If a question specifies that the equation be written in slope-intercept form, the equation should be manipulated to isolate *y*:

Solve: $y - (-2) = 3(x - 4)$

Distribute: $y + 2 = 3x - 12$

Subtract 2 from both sides: $y = 3x - 14$

Given Two Ordered Pairs
Given two ordered pairs for a function, (x_1, y_1) and (x_2, y_2), it is possible to determine the rate of change between the variables (slope of the line). To calculate the slope of the line, m, the values for the ordered pairs should be substituted into the formula: $m = \frac{y_2 - y_1}{x_2 - x_1}$. The expression is substituted to obtain a whole number or fraction for the slope. Once the slope is calculated, the slope and either of the ordered pairs should be substituted into the point-slope form to obtain the equation of the line.

Exponential Functions

An *exponential function* is a function of the form $f(x) = b^x$, where *b* is a positive real number other than 1. In such a function, *b* is called the *base*.

The *domain* of an exponential function is all real numbers, and the *range* is all positive real numbers. There will always be a horizontal asymptote of $y = 0$ on one side. If *b* is greater than 1, then the graph will be increasing moving to the right. If *b* is less than 1, then the graph will be decreasing moving to the right. Exponential functions are one-to-one. The basic exponential function graph will go through the point (0,1).

Example
Solve $5^{x+1} = 25$.

Get the x out of the exponent by rewriting the equation $5^{x+1} = 5^2$ so that both sides have a base of 5.

Since the bases are the same, the exponents must be equal to each other.

This leaves $x + 1 = 2$ or $x = 1$.

To check the answer, the x-value of 1 can be substituted back into the original equation.

Logarithmic Functions

A *logarithmic function* is an inverse for an exponential function. The inverse of the base *b* exponential function is written as $\log_b(x)$, and is called the *base b logarithm*. The domain of a logarithm is all positive real numbers. It has the properties that $\log_b(b^x) = x$. For positive real values of *x*, $b^{\log_b(x)} = x$.

When there is no chance of confusion, the parentheses are sometimes skipped for logarithmic functions: $\log_b(x)$ may be written as $\log_b x$. For the special number *e*, the base *e* logarithm is called the *natural logarithm* and is written as $\ln x$. Logarithms are one-to-one.

When working with logarithmic functions, it is important to remember the following properties. Each one can be derived from the definition of the logarithm as the inverse to an exponential function:

$$\log_b 1 = 0$$

$$\log_b b = 1$$

$$\log_b b^p = p$$

$$\log_b MN = \log_b M + \log_b N$$

$$\log_b \frac{M}{N} = \log_b M - \log_b N$$

$$\log_b M^p = p \log_b M$$

When solving equations involving exponentials and logarithms, the following fact should be used:

If *f* is a one-to-one function, $a = b$ is equivalent to $f(a) = f(b)$.

Using this, together with the fact that logarithms and exponentials are inverses, allows manipulations of the equations to isolate the variable.

Example
Solve $4 = \ln(x - 4)$.

Using the definition of a logarithm, the equation can be changed to $e^4 = e^{\ln(x-4)}$.

The functions on the right side cancel with a result of $e^4 = x - 4$.

This then gives $x = 4 + e^4$.

Trigonometric Functions

Trigonometric functions are built out of two basic functions, the *sine* and *cosine*, written as $\sin\theta$ and $\cos\theta$ respectively. Note that similar to logarithms, it is customary to drop the parentheses as long as the result is not confusing.

The sine and cosine are defined using the *unit circle*. If θ is the angle going counterclockwise around the origin from the *x*-axis, then the point on the unit circle in that direction will have the coordinates $(\cos\theta, \sin\theta)$.

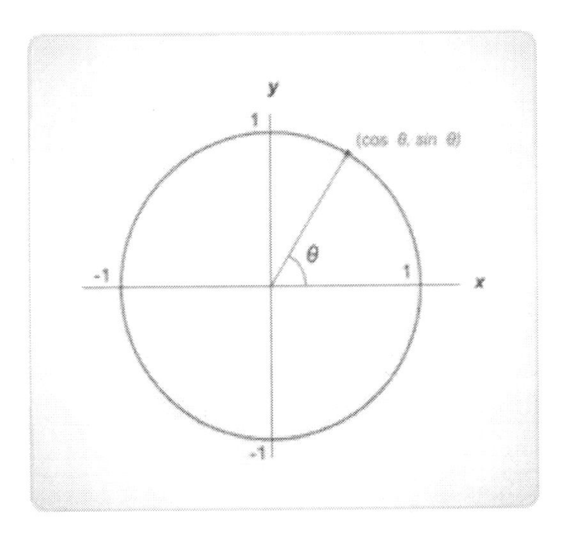

Since the angle returns to the start every 2π radians (or 360 degrees), the graph of these functions will be *periodic*, with period 2π. This means that the graph repeats itself as one moves along the *x*-axis because $\sin\theta = \sin(\theta + 2\pi)$. Cosine is works similarly.

From the unit circle definition, the sine function starts at 0 when $\theta = 0$. It grows to 1 as θ grows to $\pi/2$, and then back to 0 at $\theta = \pi$. Then it decreases to -1 as θ grows to $3\pi/2$, and back up to 0 at $\theta = 2\pi$.

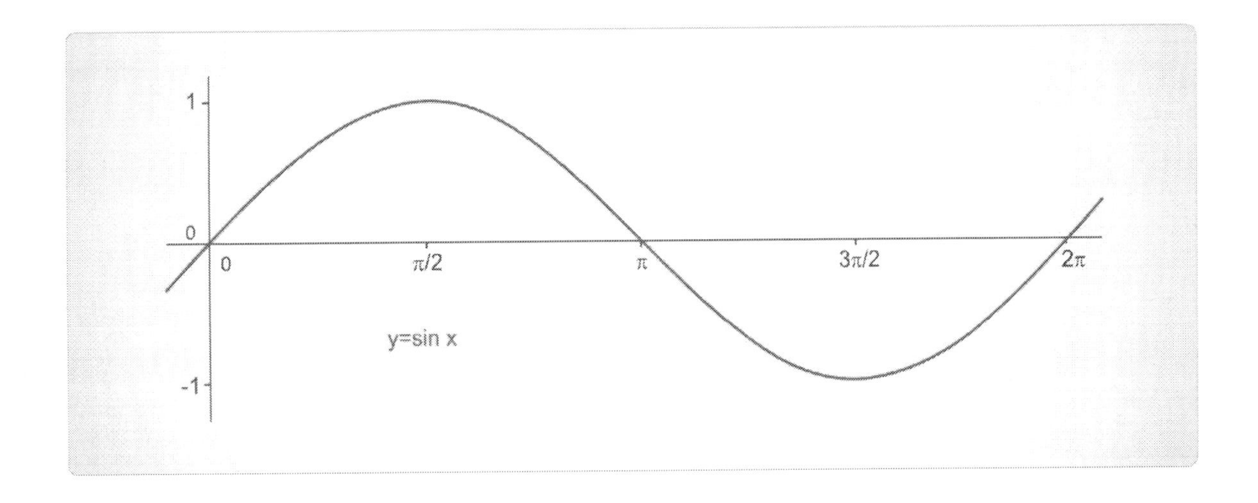

The graph of the cosine is similar. The cosine will start at 1, decreasing to 0 at $\pi/2$ and continuing to decrease to -1 at $\theta = \pi$. Then, it grows to 0 as θ grows to $3\pi/2$ and back up to 1 at $\theta = 2\pi$.

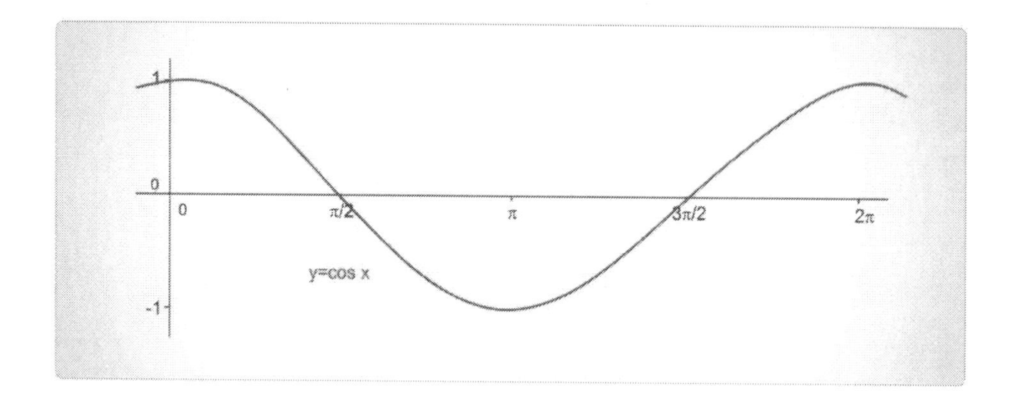

Another trigonometric function, which is frequently used, is the *tangent* function. This is defined as the following equation: $\tan\theta = \frac{\sin\theta}{\cos\theta}$.

The tangent function is a period of π rather than 2π because the sine and cosine functions have the same absolute values after a change in the angle of π, but flip their signs. Since the tangent is a ratio of the two functions, the changes in signs cancel.

The tangent function will be zero when the sine is zero, and it will have a vertical asymptote whenever cosine is zero. The following graph shows the tangent function:

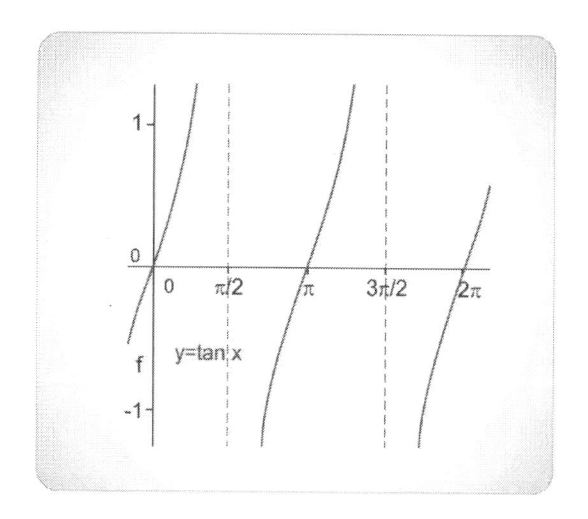

Three other trigonometric functions are sometimes useful. These are the *reciprocal* trigonometric functions, so named because they are just the reciprocals of sine, cosine, and tangent. They are the *cosecant*, defined as $\csc\theta = \frac{1}{\sin\theta}$, the *secant*, $\sec\theta = \frac{1}{\cos\theta}$, and the *cotangent*, $\cot\theta = \frac{1}{\tan\theta}$. Note that from the definition of tangent, $\cot\theta = \frac{\cos\theta}{\sin\theta}$.

In addition, there are three identities that relate the trigonometric functions to one another:

$$\cos\theta = \sin(\frac{\pi}{2} - \theta)$$

$$\csc\theta = \sec\left(\frac{\pi}{2} - \theta\right)$$

$$\cot\theta = \tan(\frac{\pi}{2} - \theta)$$

Here is a list of commonly-needed values for trigonometric functions, given in radians, for the first quadrant:

Table for trigonometric functions

$\sin 0 = 0$	$\cos 0 = 1$	$\tan 0 = 0$
$\sin\frac{\pi}{6} = \frac{1}{2}$	$\cos\frac{\pi}{6} = \frac{\sqrt{3}}{2}$	$\tan\frac{\pi}{6} = \frac{\sqrt{3}}{3}$
$\sin\frac{\pi}{4} = \frac{\sqrt{2}}{2}$	$\cos\frac{\pi}{4} = \frac{\sqrt{2}}{2}$	$\tan\frac{\pi}{4} = 1$
$\sin\frac{\pi}{3} = \frac{\sqrt{3}}{2}$	$\cos\frac{\pi}{3} = \frac{1}{2}$	$\tan\frac{\pi}{3} = \sqrt{3}$
$\sin\frac{\pi}{2} = 1$	$\cos\frac{\pi}{2} = 0$	$\tan\frac{\pi}{2} = undefined$
$\csc 0 = undefined$	$\sec 0 = 1$	$\cot 0 = undefined$
$\csc\frac{\pi}{6} = 2$	$\sec\frac{\pi}{6} = \frac{2\sqrt{3}}{3}$	$\cot\frac{\pi}{6} = \sqrt{3}$
$\csc\frac{\pi}{4} = \sqrt{2}$	$\sec\frac{\pi}{4} = \sqrt{2}$	$\cot\frac{\pi}{4} = 1$
$\csc\frac{\pi}{3} = \frac{2\sqrt{3}}{3}$	$\sec\frac{\pi}{3} = 2$	$\cot\frac{\pi}{3} = \frac{\sqrt{3}}{3}$
$\csc\frac{\pi}{2} = 1$	$\sec\frac{\pi}{2} = undefined$	$\cot\frac{\pi}{2} = 0$

To find the trigonometric values in other quadrants, complementary angles can be used. The *complementary angle* is the smallest angle between the x-axis and the given angle.

Once the complementary angle is known, the following rule is used:

For an angle θ with complementary angle x, the absolute value of a trigonometry function evaluated at θ is the same as the absolute value when evaluated at x.

The correct sign is used based on the functions sine and cosine are given by the *x* and *y* coordinates on the unit circle.

Sine will be positive in quadrants I and II and negative in quadrants III and IV.

Cosine will be positive in quadrants I and IV, and negative in II and III.

Tangent will be positive in I and III, and negative in II and IV.

The signs of the reciprocal functions will be the same as the sign of the function of which they are a reciprocal.

Example
Find $\sin\frac{3\pi}{4}$.

First, the complementary angle must be found.

This angle is in the II quadrant, and the angle between it and the *x*-axis is $\frac{\pi}{4}$.

Now, $\sin\frac{\pi}{4} = \frac{\sqrt{2}}{2}$.

Since this is in the II quadrant, sine takes on positive values (the *y* coordinate is positive in the II quadrant).

Therefore, $\sin\frac{3\pi}{4} = \frac{\sqrt{2}}{2}$.

In addition to the six trigonometric functions defined above, there are inverses for these functions. However, since the trigonometric functions are not one-to-one, one can only construct inverses for them on a restricted domain.

Usually, the domain chosen will be $[0, \pi)$ for cosine and $(-\frac{\pi}{2}, \frac{\pi}{2}]$ for sine. The inverse for tangent can use either of these domains. The inverse functions for the trigonometric functions are also called *arc functions*. In addition to being written with a -1 in the exponent to denote that the function is an inverse, they will sometimes be written with an "a" or "arc" in front of the function name, so $\cos^{-1}\theta = \operatorname{acos}\theta = \arccos\theta$.

When solving equations that involve trigonometric functions, there are often multiple solutions. For example, $2\sin\theta = \sqrt{2}$ can be simplified to $\sin\theta = \frac{\sqrt{2}}{2}$. This has solutions $\theta = \frac{\pi}{4}, \frac{3\pi}{4}$, but in addition, because of the periodicity, any integer multiple of 2π can also be added to these solutions to find another solution.

The full set of solutions is $\theta = \frac{\pi}{4} + 2\pi k, \frac{3\pi}{4} + 2\pi k$ for all integer values of k. It is very important to remember to find all possible solutions when dealing with equations that involve trigonometric functions.

The name *trigonometric* comes from the fact that these functions play an important role in the geometry of triangles, particularly right triangles.

Consider the right triangle shown in this figure:

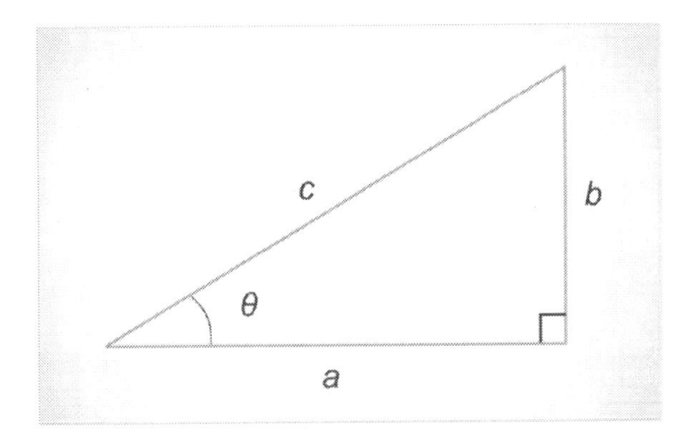

The following hold true:

- $c \sin\theta = b$
- $c \cos\theta = a$
- $\tan\theta = \frac{b}{a}$
- $b \csc\theta = c$
- $a \sec\theta = c$
- $\cot\theta = \frac{a}{b}$

Remember also the angles of a triangle must add up to π radians (180 degrees).

Practice Questions

1. Find the determinant of the matrix $\begin{bmatrix} -4 & 2 \\ 3 & -1 \end{bmatrix}$.

   a. -10
   b. -2
   c. 0
   d. 2

2. If $a \neq b$, solve for x if $\frac{1}{x} + \frac{2}{a} = \frac{2}{b}$

   a. $\frac{a-b}{ab}$

   b. $\frac{ab}{2(a-b)}$

   c. $\frac{2(a-b)}{ab}$

   d. $\frac{a-b}{2ab}$

3. If $x^2 + x - 3 = 0$, then $\left(x - \frac{1}{2}\right)^2 =$

   a. $\frac{11}{2}$

   b. $\frac{11}{4}$

   c. 11

   d. $\frac{121}{4}$

4. Which graph will be a line parallel to the graph of $y = 3x - 2$?
   a. $2y - 6x = 2$
   b. $y - 4x = 4$
   c. $3y = x - 2$
   d. $2x - 2y = 2$

5. An equation for the line passing through the origin and the point $(2, 1)$ is
   a. $y = 2x$
   b. $y = \frac{1}{2}x$
   c. $y = x - 2$
   d. $2y = x + 1$

6. Jessica buys 10 cans of paint. Red paint costs \$1 per can and blue paint costs \$2 per can. In total, she spends \$16. How many red cans did she buy?
   a. 2
   b. 3
   c. 4
   d. 5

7. A farmer owns two (non-adjacent) square plots of land, which he wishes to fence. The area of one is 1000 square feet, while the area of the other is 10 square feet. How much fencing does he need, in feet?

a. 44

b. $40\sqrt{10}$

c. $440\sqrt{10}$

d. $44\sqrt{10}$

8. If $\log_{10} x = 2$, then *x* is

a. 4

b. 20

c. 100

d. 1000

9. Let $f(x) = 2x + 1, g(x) = \frac{x-1}{4}$. Find $g(f(x))$.

a. $\frac{x+1}{2}$

b. $\frac{x}{2}$

c. $\frac{2x^2-x-1}{4}$

d. $3x$

10. Suppose θ is an acute angle, and $\sin\theta = \frac{\sqrt{3}}{2}$. What is $\cos\theta$?

a. $\frac{1}{2}$

b. $\frac{\sqrt{3}}{2}$

c. $\frac{\sqrt{2}}{2}$

d. $\frac{1}{4}$

11. $2x(3x + 1) - 5(3x + 1) =$

a. $10x(3x + 1)$

b. $10x^2(3x + 1)$

c. $(2x - 5)(3x + 1)$

d. $(2x + 1)(3x - 5)$

12. For which real numbers *x* is $-3x^2 + x - 8 > 0$?

a. All real numbers *x*

b. $-2\sqrt{\frac{2}{3}} < x < 2\sqrt{\frac{2}{3}}$

c. $1 - 2\sqrt{\frac{2}{3}} < x < 1 + 2\sqrt{\frac{2}{3}}$

d. For no real numbers *x*

13. A root of $x^2 - 2x - 2$ is

a. $1 + \sqrt{3}$

b. $1 + 2\sqrt{2}$

c. $2 + 2\sqrt{3}$

d. $2 - 2\sqrt{3}$

14. In the *xy*-plane, the graph of $y = x^2 + 2$ and the circle with center $(0,1)$ and radius 1 have how many points of intersection?

a. 0

b. 1

c. 2

d. 3

15. A line goes through the point (-4, 0) and the point (0,2). What is the slope of the line?

a. 2

b. 4

c. $\frac{3}{2}$

d. $\frac{1}{2}$

16. How many different ways can we order the letters *a, b, c*?

a. 3

b. 6

c. 9

d. 12

17. If $f(x) = 4x + 2$, and $f^{-1}(x)$ is the inverse function for f, then what is $f^{-1}(6)$?

a. 0

b. $\frac{1}{2}$

c. 1

d. $\frac{3}{2}$

18. The sequence $\{a_n\}$ is defined by the relation $a_{n+1} = 3a_n - 1, a_1 = 1$. Find a_3.

a. 2

b. 4

c. 5

d. 15

19. Six people apply to work for Janice's company, but she only needs four workers. How many different groups of four employees can Janice choose?

a. 6

b. 10

c. 15

d. 36

20. If $f(x) = (\frac{1}{2})^x$ and $a < b$, then which of the following must be true?

a. $f(a) < f(b)$

b. $f(a) > f(b)$

c. $f(a) + f(b) = 0$

d. $3f(a) = f(b)$

21. For the following similar triangles, what are the values of x and y (rounded to one decimal place)?

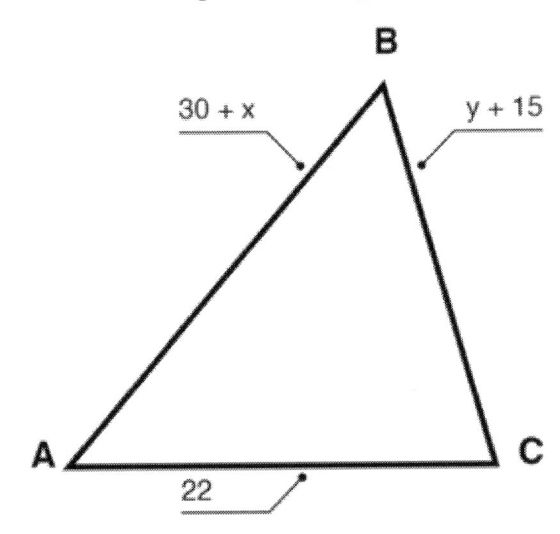

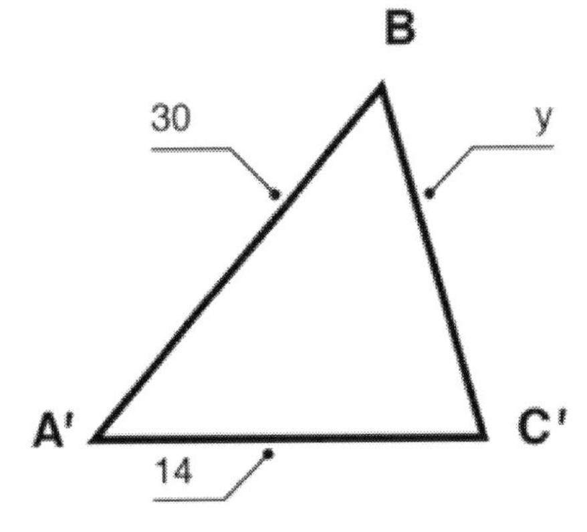

a. $x = 16.5, y = 25.1$
b. $x = 19.5, y = 24.1$
c. $x = 17.1, y = 26.3$
d. $x = 26.3, y = 17.1$

22. The triangle shown below is a right triangle. What's the value of x?

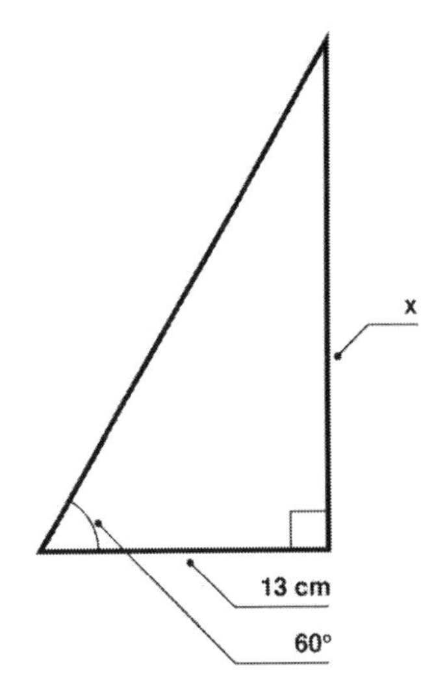

a. $x = 1.73$
b. $x = 0.57$
c. $x = 13$
d. $x = 22.49$

23. Which of the following shows a line of symmetry?

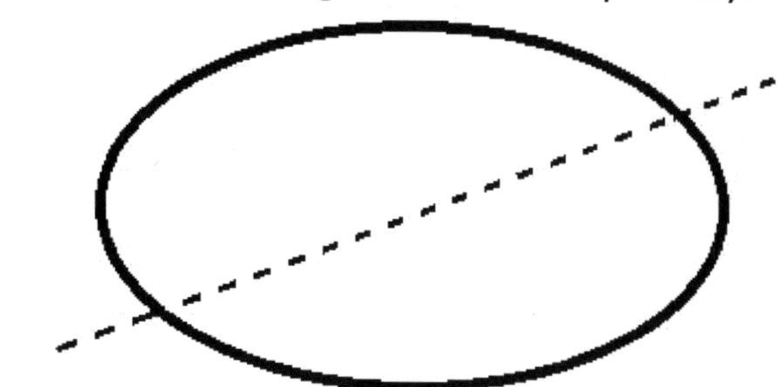

a.

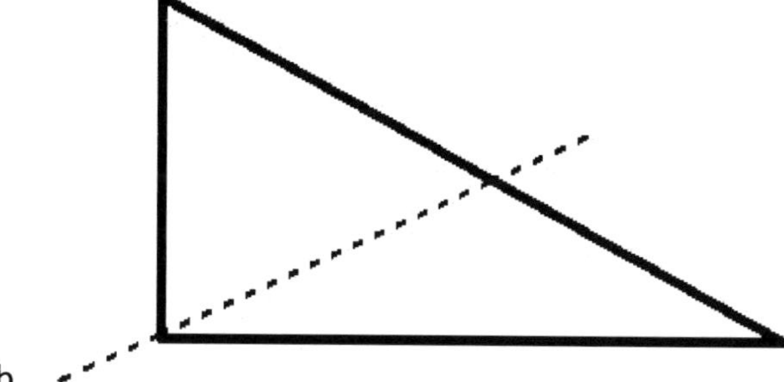

b.

c.

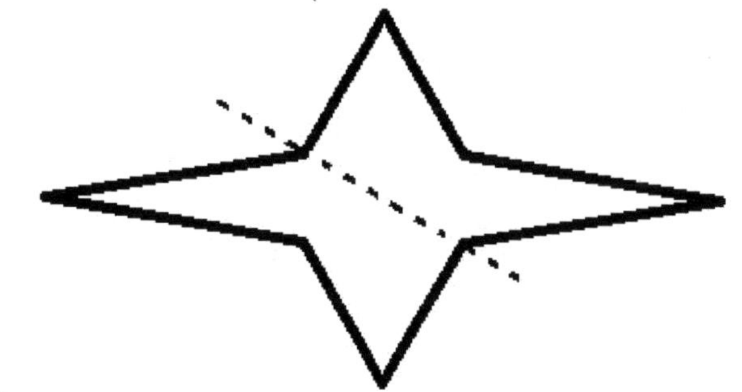

d.

24. What are the coordinates of the focus of the parabola $y = -9x^2$?

a. $(-3, 0)$

b. $\left(-\frac{1}{36}, 0\right)$

c. $(0, -3)$

d. $\left(0, -\frac{1}{36}\right)$

25. If $4x - 3 = 5$, then $x =$

a. 1

b. 2

c. 3

d. 4

Answer Explanations

1. B: The determinant of a 2 x 2 matrix is $ad - bc$. The calculation is -4(-1) – 2(3) = 4 – 6 = -2.

2. B: $\frac{2}{a}$ must be subtracted from both sides, with a result of $\frac{1}{x} = \frac{2}{b} - \frac{2}{a}$. The reciprocal of both sides needs to be taken, but the right-hand side needs to be written as a single fraction in order to do that. Since the two fractions on the right have denominators that are not equal, a common denominator of *ab* is needed. This leaves $\frac{1}{x} = \frac{2a}{ab} - \frac{2b}{ab} = \frac{2(a-b)}{ab}$. Taking the reciprocals, which can be done since *b* – *a* is not zero, with a result of $x = \frac{ab}{2(a-b)}$.

3. B: Plugging into the quadratic formula yields, for solutions $x = \frac{1 \pm \sqrt{-1+4 \cdot 1(-3)}}{2} = \frac{1}{2} \pm \frac{\sqrt{11}}{2}$. Therefore, $x - \frac{1}{2} = \pm \frac{\sqrt{11}}{2}$. Now, if this is squared, then the $\pm$ cancels and left with $\left(\frac{\sqrt{11}}{2}\right)^2 = \frac{11}{4}$.

4. A: Parallel lines have the same slope. The slope of *C* can be seen to be 1/3 by dividing both sides by 3. For *E*, divide both sides by 2 and see the slope is 3/2. The others are in standard form $Ax + By = C$, for which the slope is given by $\frac{-A}{B}$. The slope of *A* is 3, the slope of *B* is 4. The slope of *D* is 1.

5. B: The slope will be given by $\frac{1-0}{2-0} = \frac{1}{2}$. The *y*-intercept will be 0, since it passes through the origin. Using slope-intercept form, the equation for this line is $y = \frac{1}{2}x$.

6. C: Let *r* be the number of red cans and *b* be the number of blue cans. One equation is $r + b = 10$. The total price is $16, and the prices for each can means $1r + 2b = 16$. Multiplying the first equation on both sides by -1 results in $-r - b = -10$. Add this equation to the second equation, leaving $b = 6$. So, she bought 6 *blue* cans. From the first equation, this means *r* = 4; thus, she bought 4 *red* cans.

7. D: The first field has an area of 1000 feet, so the length of one side is $\sqrt{1000} = 10\sqrt{10}$. Since there are four sides to a square, the total perimeter is $40\sqrt{10}$. The second square has an area of 10 square feet, so the length of one side is $\sqrt{10}$, and the total perimeter is $4\sqrt{10}$. Adding these together gives $40\sqrt{10} + 4\sqrt{10} = (40 + 4)\sqrt{10} = 44\sqrt{10}$.

8. C: If $\log_{10} x = 2$, then $10^2 = x$, which equals 100.

9. B: Recall that to compose functions, replace the *x* in the expression for *g* with the expression for *f* everywhere there is x. So $g(f(x)) = \frac{f(x)-1}{4} = \frac{2x+1-1}{4} = \frac{2x}{4} = \frac{x}{2}$.

10. A: For acute angles, the only angle for which $\sin\theta = \frac{\sqrt{3}}{2}$ is $\frac{\pi}{3}$. Also, $\cos\frac{\pi}{3} = \frac{1}{2}$.

11. C: The $(3x + 1)$ can be factored to get $(2x - 5)(3x + 1)$.

12. D: Because the coefficient of x^2 is negative, this function has a graph that is a parabola that opens downward. Therefore, it will be greater than 0 between its real roots, if it has any. Checking the discriminant, the result is $1^2 - 4(-3)(-8) = 1 - 96 = -95$. Since the discriminant is negative, this equation has no real solutions. Since this has no real roots, it must be always positive or always negative. Its graph opens downward, so it has at least some negative values. That means it is always negative. Thus, it is bigger than zero for no real numbers.

13. A: Check each value, but it is easiest to use the quadratic formula, which gives $x = \frac{2 \pm \sqrt{(-2)^2 - 4(1)(-2)}}{2} = 1 \pm \frac{\sqrt{12}}{2} = 1 \pm \frac{2\sqrt{3}}{2} = 1 \pm \sqrt{3}$. The only one of these which appears as an answer choice is $1 + \sqrt{3}$.

14. B: The *y* coordinate of every point on the graph of $y = x^2 + 2$ has a vertex at (0,2) on the y-axis. The circle with a center at (0,1) also lies on the y-axis. With a radius of 1, the circle touches the parabola at one point. The vertex of the parabola (0,2).

15. D: The slope is given by the change in *y* divided by the change in *x*. The change in *y* is 2-0 = 2, and the change in *x* is 0 – (-4) = 4. The slope is $\frac{2}{4} = \frac{1}{2}$.

16. B: The number of ways to order *n* objects is given by the product $\binom{n}{1}\binom{n-1}{1} \ldots \binom{1}{1}$. This is $\binom{3}{1}\binom{2}{1}\binom{1}{1} = \frac{3!}{1!2!} \cdot \frac{2!}{1!1!} \cdot \frac{1!}{1!1!}$. 1! is just 1, and the 2! in the numerator and denominator will cancel one another, with a result of $3! = 3 \cdot 2 \cdot 1 = 6$.

17. C: If $y = f^{-1}(6)$ then *y* must satisfy $f(y) = 6$. Substituting and solving for y yields $4y + 2 = 6$, then $4y = 4$, and $y = 1$.

18. C: Find $a_2 = 3a_1 - 1 = 3 \cdot 1 - 1 = 2$. Next, find $a_3 = 3a_2 - 1 = 3 \cdot 2 - 1 = 5$.

19. C: Janice will be choosing 4 employees out of a set of 6 applicants, so this will be given by the choice function. The following equation shows the choice function worked out: $\binom{6}{4} = \frac{6!}{4!(6-4)!} = \frac{6!}{4!(2)!} = \frac{6 \cdot 5 \cdot 4 \cdot 3 \cdot 2 \cdot 1}{4 \cdot 3 \cdot 2 \cdot 1 \cdot 2 \cdot 1} = \frac{6 \cdot 5}{2} = 15$.

20. B: Here, *f* is an exponential function whose base is less than 1. In this function, *f* is always decreasing. This means that when *a* is less than *b*, $f(a) > f(b)$.

21. C: Because the triangles are similar, the lengths of the corresponding sides are proportional. Therefore:

$$\frac{30 + x}{30} = \frac{22}{14} = \frac{y + 5}{y}$$

This results in the equation $14(30 + x) = 22 \cdot 30$ which, when solved, gives $x = 17.1$. The proportion also results in the equation $14(y + 5) = 22y$ which, when solved, gives $y = 26.3$.

21. D: SOHCAHTOA is used to find the missing side length. Because the angle and adjacent side are known, $\tan 60 = \frac{x}{13}$. Making sure to evaluate tangent with an argument in degrees, this equation gives:

$$x = 13 \tan 60 = 13 \cdot 1.73 = 22.49$$

23. C: The triangle in Choice B doesn't contain a line of symmetry. The figures in Choices A and D do contain a line of symmetry but it is not the line that is shown here. Choice C is the only one with a correct line of symmetry shown, such that the figure is mirrored on each side of the line.

24. D: A parabola of the form $y = \frac{1}{4f}x^2$ has a focus $(0, f)$. Because $y = -9x^2$, set $-9 = \frac{1}{4f}$. Solving this equation for f results in $f = -\frac{1}{36}$. Therefore, the coordinates of the focus are $\left(0, -\frac{1}{36}\right)$.

25. B: Add 3 to both sides to get $4x = 8$. Then divide both sides by 4 to get $x = 2$.

Elementary Algebra

Operations with Integers and Rational Numbers

Computation with Integers and Negative Rational Numbers

Integers are the whole numbers together with their negatives. They include numbers like 5, 24, 0, -6, and 15. They do not include fractions or numbers that have digits after the decimal point.

Rational numbers are all numbers that can be written as a fraction using integers. A *fraction* is written as $\frac{x}{y}$ and represents the quotient of *x* being divided by *y*. More practically, it means dividing the whole into *y* equal parts, then taking *x* of those parts.

Examples of rational numbers include $\frac{1}{2}$ and $\frac{5}{4}$. The number on the top is called the *numerator*, and the number on the bottom is called the *denominator*. Because every integer can be written as a fraction with a denominator of 1, (e.g. $\frac{3}{1} = 3$), every integer is also a rational number.

When adding integers and negative rational numbers, there are some basic rules to determine if the solution is negative or positive:

Adding two positive numbers results in a positive number: 3.3 + 4.8 = 8.1.

Adding two negative numbers results in a negative number: (-8) + (-6) = -14.

Adding one positive and one negative number requires taking the absolute values and finding the difference between them. Then, the sign of the number that has the higher absolute value for the final solution is used.

For example, (-9) + 11, has a difference of absolute values of 2. The final solution is 2 because 11 has the higher absolute value. Another example is 9 + (-11), which has a difference of absolute values of 2. The final solution is -2 because 11 has the higher absolute value.

When subtracting integers and negative rational numbers, one has to change the problem to adding the opposite and then apply the rules of addition.

Subtracting two positive numbers is the same as adding one positive and one negative number.

For example, 4.9 – 7.1 is the same as 4.9 + (-7.1). The solution is -2.2 since the absolute value of -7.1 is greater. Another example is 8.5 – 6.4 which is the same as 8.5 + (-6.4). The solution is 2.1 since the absolute value of 8.5 is greater.

Subtracting a positive number from a negative number results in negative value.

For example, (-12) – 7 is the same as (-12) + (-7) with a solution of -19.

Subtracting a negative number from a positive number results in a positive value.

For example, 12 – (-7) is the same as 12 + 7 with a solution of 19.

For multiplication and division of integers and rational numbers, if both numbers are positive or both numbers are negative, the result is a positive value.

For example, (-1.7)(-4) has a solution of 6.8 since both numbers are negative values.

If one number is positive and another number is negative, the result is a negative value.

For example, (-15)/5 has a solution of -3 since there is one negative number.

The Use of Absolute Values

The *absolute value* represents the distance a number is from 0. The *absolute value symbol* is | | with a number between the bars. The |10| = 10 and the |-10| = 10.

When simplifying an algebraic expression, the value of the absolute value expression is determined first, much like parenthesis in the order of operations. See the example below:

$$|8 - 12| + 5 = |-4| + 5 = 4 + 5 = 9$$

Order of Operations

Exponents are shorthand for longer multiplications or divisions. The exponent is written to the upper right of a number. In the expression 2^3, the exponent is 3. The number with the exponent is called the *base*.

When the exponent is a whole number, it means to multiply the base by itself as many times as the number in the exponent. So, $2^3 = 2 \times 2 \times 2 = 8$.

If the exponent is a negative number, it means to take the reciprocal of the positive exponent:

$$2^{-3} = \frac{1}{2^3} = \frac{1}{8}$$

When the exponent is 0, the result is always 1: $2^0 = 1, 5^0 = 1$, and so on.

When the exponent is 2, the number is *squared*, and when the exponent is 3, it is *cubed*.

When working with longer expressions, parentheses are used to show the order in which the operations should be performed. Operations inside the parentheses should be completed first. Thus, $(3 - 1) \div 2$ means one should first subtract 1 from 3, and then divide that result by 2.

The *order of operations* gives an order for how a mathematical expression is to be simplified:

- Parentheses
- Exponents
- Multiplication
- Division
- Addition
- Subtraction

To help remember this, many students like to use the mnemonic PEMDAS. Some students associate this word with a phrase to help them, such as "Pirates Eat Many Donuts at Sea." Here is a quick example:

Evaluate $2^2 \times (3 - 1) \div 2 + 3$.

Parenthesis: $2^2 \times 2 \div 2 + 3$.

Exponents: $4 \times 2 \div 2 + 3$

Multiply: $8 \div 2 + 3$.

Divide: $4 + 3$.

Addition: 7

Ordering of Numbers

In counting, when a number appears after another number in order, that number will be one more. On the other hand, when a number appears before another number in order, that number will be one less. This idea is useful when counting backward. Also, zero means that there is none of something. This idea can be seen by taking away all of something so that there are zero items left. Also, learning to count by tens starting at any number is a key concept. Once a new number is learned, learning how to read and write that number is also important.

Placing numbers in an order in which they are listed from smallest to largest is known as *ordering*. When items are listed by using numbers in order, the *ordinal numbers*, 1st, 2nd, 3rd, 4th, ..., can be used.

When you order numbers the right way, you can more easily compare the different amounts of items. When you compare numbers you show whether two amounts are the same or different. Teachers can show two different quantities of items in the classroom. Then they can discuss which amount is lesser or greater. This exercise also can be used in order to classify numbers from the smallest amount to the largest amount.

Being able to compare any two whole numbers without a visual representation is also an important task. Each whole number relates to a certain amount. This amount can be ranked and compared to other amounts. Knowing the right vocabulary relating to ordering and comparing is important. The *equals sign* is $=$. It shows that two numbers are the same on either side of the symbol. For example, $28 = 28$. The symbols that are used for comparison are $<$ to represent *less than*, $>$ to represent *greater than*. The symbols $\leq$ to represent *less than or equal to*, and $\geq$ to represent *greater than or equal to*, and $\neq$ to represent *not equal to* can also be used.

You can compare numbers with any number of digits when you use these symbols. For example, the expression $77 < 100$, should be understood as 77 is less than 100. The expression $44 > 23$ should be understood as 44 is greater than 23. The expression $22 \neq 24$ should be understood as 22 is not equal *to 24*. Also, both $36 = 36$ and $36 \leq 36$ can be written because both "36 equals 36" and "36 is less than or equal to 36" applies.

Patterns

Patterns are an important part of mathematics. When mathematical calculations are completed repeatedly, patterns can be recognized. Recognizing patterns is an integral part of mathematics because it helps you understand relationships between different ideas. For example, a sequence of numbers can be given, and being able to recognize the relationship between the given numbers can help in completing the sequence.

For instance, given the sequence of numbers $7, 14, 21, 28, 35, \ldots$, the next number in the sequence would be 42. This is because the sequence lists all multiples of 7, starting at 7. Sequences can also be built from addition, subtraction, and division. Being able to recognize the relationship between the values that are given is the key to finding out the next number in the sequence.

Patterns within a sequence can come in 2 distinct forms. The items either repeat in a constant order, or the items change from one step to another in some consistent way. The core is the smallest unit, or number of items, that repeats in a repeating pattern. For example, the pattern oo▲oo▲o... has a core that is oo▲. Knowing only the core, the pattern can be extended. Knowing the number of steps in the core allows the identification of an item in each step without drawing/writing the entire pattern out. For example, suppose you must find the tenth item in the previous pattern. Because the core consists of three items (oo▲), the core repeats in multiples of 3. In other words, steps 3, 6, 9, 12, etc. will be ▲ completing the core with the core starting over on the next step. For the above example, the 9th step will be ▲ and the 10th will be o.

The most common patterns where each item changes from one step to the next are arithmetic and geometric sequences. In an arithmetic sequence, the items increase or decrease by a constant difference. In other words, the same thing is added or subtracted to each item or step to produce the next. To determine if a sequence is arithmetic, see what must be added or subtracted to step one to produce step two. Then, check if the same thing is added/subtracted to step two to produce step three. The same thing must be added/subtracted to step three to produce step four, and so on. Consider the pattern 13, 10, 7, 4, ... To get from step one (13) to step two (10) by adding or subtracting requires subtracting by 3. The next step is checking if subtracting 3 from step two (10) will produce step three (7), and subtracting 3 from step three (7) will produce step four (4). In this case, the pattern holds true. Therefore, this is an arithmetic sequence in which each step is produced by subtracting 3 from the previous step. To extend the sequence, 3 is subtracted from the last step to produce the next. The next three numbers in the sequence are 1, -2, -5.

A geometric sequence is one in which each step is produced by multiplying or dividing the previous step by the same number. To see if a sequence is geometric, decide what step one must be multiplied or divided by to produce step two. Then check if multiplying or dividing step two by the same number produces step three, and so on. Consider the pattern 2, 8, 32, 128, ... To get from step one (2) to step two (8) requires multiplication by 4. The next step determines if multiplying step two (8) by 4 produces step three (32), and multiplying step three (32) by 4 produces step four (128). In this case, the pattern holds true. Therefore, this is a geometric sequence in which each step is found by multiplying the previous step by 4. To extend the sequence, the last step is multiplied by 4 and repeated. The next three numbers in the sequence are 512; 2,048; 8,192.

Arithmetic and geometric sequences can also be represented by shapes. For example, an arithmetic sequence could consist of shapes with three sides, four sides, and five sides. A geometric sequence could consist of eight blocks, four blocks, and two blocks (each step is produced by dividing the number of blocks in the previous step by 2).

Relationships Between the Corresponding Terms of Two Numerical Patterns

When given two number patterns, the corresponding terms should be examined to determine if a relationship exists between them. Corresponding terms between patterns are the pairs of numbers which appear in the same step of the two sequences. Consider the following patterns 1, 2, 3, 4,... and 3, 6, 9, 12, ... The corresponding terms are: 1 and 3; 2 and 6; 3 and 9; and 4 and 12. To identify the relationship, each pair of corresponding terms is examined. You can also examine the possibilities of performing an operation (+, −, ×, ÷) to each sequence. In this case:

$1 + 2 = 3$ or $1 \times 3 = 3$

$2 + 4 = 6$ or $2 \times 3 = 6$

$3 + 6 = 9$ or $3 \times 3 = 9$

$4 + 8 = 12$ or $4 \times 3 = 12$

The pattern is that the number from the first sequence multiplied by 3 equals the number in the second sequence. By assigning each sequence a label (input and output) or variable (*x* and *y*), the relationship can be written as an equation. The first sequence represents the inputs, or *x*, and the second sequence represents the outputs, or *y*. So, the relationship can be expressed as: $y = 3x$.

Consider the following sets of numbers:

a	2	4	6	8
b	6	8	10	12

To write a rule for the relationship between the values for *a* and the values for *b*, the corresponding terms (2 and 6; 4 and 8; 6 and 10; 8 and 12) are examined. The possibilities for producing *b* from *a* are:

$2 + 4 = 6$ or $2 \times 3 = 6$

$4 + 4 = 8$ or $4 \times 2 = 8$

$6 + 4 = 10$

$8 + 4 = 12$ or $8 \times 1.5 = 12$

The pattern is that adding 4 to the value of *a* produces the value of *b*. The relationship can be written as the equation $a + 4 = b$.

Operations with Algebraic Expressions

Evaluation of Simple Formulas and Expressions

To evaluate simple formulas and expressions, the first step is to substitute the given values in for the variable(s). Then, the order of operations is used to simplify.

Example 1

Evaluate $\frac{1}{2}x^2 - 3, x = 4$.

The first step is to substitute in 4 for *x* in the expression: $\frac{1}{2}(4)^2 - 3$.

Then, the order of operations is used to simplify.

The exponent comes first, $\frac{1}{2}(16) - 3$, then the multiplication 8 – 3, and then, after subtraction, the solution is 5.

Example 2
Evaluate $4|5 - x| + 2y, x = 4, y = -3$.

The first step is to substitute 4 in for *x* and -3 in for *y* in the expression: $4|5 - 4| + 2(-3)$.

Then, the absolute value expression is simplified, which is $|5 - 4| = |1| = 1$.

The expression is $4(1) + 2(-3)$ which can be simplified using the order of operations.

First is the multiplication, 4 + (-6); then addition yields an answer of -2.

Example 3
Find the perimeter of a rectangle with a length of 6 inches and a width of 9 inches.

The first step is substituting in 6 for the length and 9 for the width in the perimeter of a rectangle formula, $P = 2(6) + 2(9)$.

Then, the order of operations is used to simplify.

First is multiplication (resulting in 12 + 21) and then addition for a solution of 33 inches.

Adding and Subtracting Monomials and Polynomials

To add or subtract polynomials, add the coefficients of terms with the same exponent. For instance, $(-2x^2 + 3x + 1) + (4x^2 - x) = (-2 + 4)x^2 + (3 - 1)x + 1 = 2x^2 + 2x + 1$.

Multiplying and Dividing Monomials and Polynomials

To multiply polynomials each term of the first polynomial multiplies each term of the second polynomial, and adds up the results. Here's an example:

$$(3x^4 + 2x^2)(2x^2 + 3) = 3x^4 \cdot 2x^2 + 3x^4 \cdot 3 + 2x^2 \cdot 2x^2 + 2x^2 \cdot 3$$

Then, add like terms with a result of:

$$6x^6 + 9x^4 + 4x^4 + 6x^2 = 6x^6 + 13x^4 + 6x^2$$

A polynomial with two terms is called a *binomial*. Another way of remember the rule for multiplying two binomials is to use the acronym *FOIL*: multiply the *F*irst terms together, then the *O*utside terms (terms on the far left and far right), then the *I*nner terms, and finally the *L*ast two terms. For longer polynomials, there is no such convenient mnemonic, so remember to multiply each term of the first polynomial by each term of the second, and add the results.

To divide one polynomial by another, the procedure is similar to long division. At each step, one needs to figure out how to get the term of the dividend with the highest exponent as a multiple of the divisor. The divisor is multiplied by the multiple to get that term, which goes in the quotient. Then, the product of this term is subtracted with the dividend from the divisor and repeat the process. This sounds rather abstract, so it may be easiest to see the procedure by describing it while looking at an example.

Example

$(4x^3 + x^2 - x + 4) \div (2x - 1)$

The first step is to cancel out the highest term in the first polynomial.

To get $4x^3$ from the second polynomial, multiply by $2x^2$.

The first term for the quotient is going to be $2x^2$.

The result of $2x^2(2x - 1)$ is $4x^3 - 2x^2$. Subtract this from the first polynomial.

The result is $(-x^2 - x + 4) \div (2x - 1)$.

The procedure is repeated: to cancel the $-x^2$ term, then multiply $(2x - 1)$ by $-\frac{1}{2}x$.

Adding this to the quotient, the quotient becomes $2x^2 - \frac{1}{2}x$.

The dividend is changed by subtracting $-\frac{1}{2}x(2x - 1)$ from it to obtain $(-\frac{3}{2}x + 4) \div (2x - 1)$.

To get $-\frac{3}{2}x$ needs to be multiplied by $-\frac{3}{4}$.

The quotient, therefore, becomes $2x^2 - \frac{1}{2}x - \frac{3}{4}$.

The remaining part is $4.75 \div (2x - 1)$.

There is no monomial to multiply to cancel this constant term, since the divisor now has a higher power than the dividend.

The final answer is the quotient plus the remainder divided by $(2x - 1)$: $2x^2 - \frac{1}{2}x - \frac{3}{4} + \frac{4.75}{2x-1}$.

Zeros of Polynomials

Finding the zeros of polynomial functions is the same process as finding the solutions of polynomial equations. These are the points at which the graph of the function crosses the x-axis. As stated previously, factors can be used to find the zeros of a polynomial function. The degree of the function shows the number of possible zeros. If the highest exponent on the independent variable is 4, then the degree is 4, and the number of possible zeros is 4. If there are complex solutions, the number of roots is less than the degree.

Given the function $y = x^2 + 7x + 6$, y can be set equal to zero, and the polynomial can be factored. The equation turns into $0 = (x + 1)(x + 6)$, where $x = -1$ and $x = -6$ are the zeros. Since this is a quadratic equation, the shape of the graph will be a parabola. Knowing that zeros represent the points where the parabola crosses the x-axis, the maximum or minimum point is the only other piece needed to sketch a rough graph of the function. By looking at the function in standard form, the coefficient of x

is positive; therefore, the parabola opens *up*. Using the zeros and the minimum, the following rough sketch of the graph can be constructed:

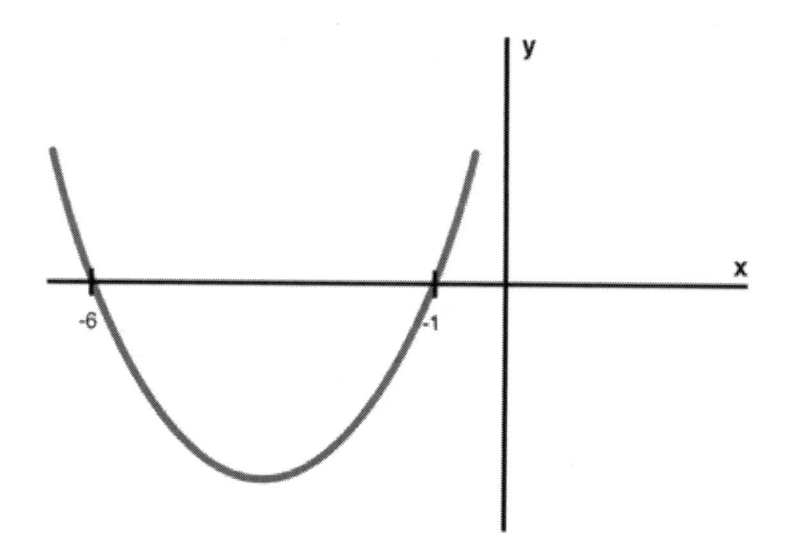

Solving Logarithmic and Exponential Functions

To solve an equation involving exponential expressions, the goal is to isolate the exponential expression. Once this process is completed, the logarithm—with the base equaling the base of the exponent of both sides—needs to be taken to get an expression for the variable. If the base is e, the natural log of both sides needs to be taken.

To solve an equation with logarithms, the given equation needs to be written in exponential form, using the fact that $\log_b a = x$ means $b^x = y$, and then solved for the given variable. Lastly, properties of logarithms can be used to simplify more than one logarithmic expression into one.

When working with logarithmic functions, it is important to remember the following properties. Each one can be derived from the definition of the logarithm as the inverse to an exponential function:

$$\log_b 1 = 0$$

$$\log_b b = 1$$

$$\log_b b^p = p$$

$$\log_b MN = \log_b M + \log_b N$$

$$\log_b \frac{M}{N} = \log_b M - \log_b N$$

$$\log_b M^p = p \log_b M$$

When solving equations involving exponentials and logarithms, the following fact should be used:

If f is a one-to-one function, $a = b$ is equivalent to $f(a) = f(b)$.

Using this, together with the fact that logarithms and exponentials are inverses, allows manipulations of the equations to isolate the variable.

Example
Solve $4 = \ln(x-4)$.

Using the definition of a logarithm, the equation can be changed to $e^4 = e^{\ln(x-4)}$.

The functions on the right side cancel with a result of $e^4 = x - 4$.

This then gives $x = 4 + e^4$.

Exponential Expressions
Exponential expressions can also be rewritten, just as quadratic equations. Properties of exponents must be understood. Multiplying two exponential expressions with the same base involves adding the exponents:

$$a^m a^n = a^{m+n}$$

Dividing two exponential expressions with the same base involves subtracting the exponents:

$$\frac{a^m}{a^n} = a^{m-n}$$

Raising an exponential expression to another exponent includes multiplying the exponents:

$$(a^m)^n = a^{mn}$$

The zero power always gives a value of 1: $a^0 = 1$. Raising either a product or a fraction to a power involves distributing that power:

$$(ab)^m = a^m b^m \text{ and } \left(\frac{a}{b}\right)^m = \frac{a^m}{b^m}$$

Finally, raising a number to a negative exponent is equivalent to the reciprocal including the positive exponent:

$$a^{-m} = \frac{1}{a^m}$$

Simplifying Algebraic Fractions

A *rational expression* is a fraction with a polynomial in the numerator and denominator. The denominator polynomial cannot be zero. An example of a rational expression is $\frac{3x^4-2}{-x+1}$. The same rules for working with addition, subtraction, multiplication, and division with rational expressions apply as when working with regular fractions.

The first step is to find a common denominator when adding or subtracting. This can be done just as with regular fractions. For example, if $\frac{a}{b} + \frac{c}{d}$, then a common denominator can be found by multiplying to find the following fractions: $\frac{ad}{bd}, \frac{cb}{db}$.

A *complex fraction* is a fraction in which the numerator and denominator are themselves fractions, of the form $\frac{(\frac{a}{b})}{(\frac{c}{d})}$. These can be simplified by following the usual rules for the order of operations, or by remembering that dividing one fraction by another is the same as multiplying by the reciprocal of the divisor. This means that any complex fraction can be rewritten using the following form: $\frac{(\frac{a}{b})}{(\frac{c}{d})} = \frac{a}{b} \cdot \frac{d}{c}$.

The following problem is an example of solving a complex fraction:

$$\frac{(\frac{5}{4})}{(\frac{3}{8})} = \frac{5}{4} \cdot \frac{8}{3} = \frac{40}{12} = \frac{10}{3}$$

Solution of Equations, Inequalities, and Word Problems

Solving Linear Equations and Inequalities

The simplest equations to solve are *linear equations*, which have the form $ax + b = 0$. These have the solution $x = -\frac{b}{a}$.

For instance, in the equation $\frac{1}{3}x - 4 = 0$, it can be determined that $\frac{1}{3}x = 4$ by adding 4 on each side. Next, both sides of the equation are multiplied by 3 to get $x = 12$.

Solving an inequality is very similar to solving equations, with one important issue to keep track of: if multiplying or dividing both sides of an inequality by a negative number, the direction of the inequality *flips*.

For example, consider the inequality $-4x < 12$. Solving this inequality requires the division of -4. When the negative four is divided, the less-than sign changes to a greater-than sign. The solution becomes $x > -3$.

Example
$-4x - 3 \leq -2x + 1$

2x is added to both sides, and 3 is added to both sides, leaving $-2x \leq 4$.

$-2x \leq 4$ is multiplied by $-\frac{1}{2}$, which means flipping the direction of the inequality.

This gives $x \geq -2$.

An *absolute inequality* is an inequality that is true for all real numbers. An inequality that is only true for some real numbers is called *conditional*.

In addition to the inequalities above, there are also *double inequalities* where three quantities are compared to one another, such as $3 \leq x + 4 < 5$. The rules for double inequalities include always performing any operations to every part of the inequality and reversing the direction of the inequality when multiplying or dividing by a negative number.

When solving equations and inequalities, the solutions can be checked by plugging the answer back in to the original problem. If the solution makes a true statement, the solution is correct.

Solving Quadratic Equations by Factoring

Solving quadratic equations is a little trickier. If they take the form $ax^2 - b = 0$, then the equation can be solved by adding *b* to both sides and dividing by *a* to get $x^2 = \frac{b}{a}$.

Using the sixth rule above, the solution is $x = \pm\sqrt{\frac{b}{a}}$. Note that this is actually two separate solutions, unless *b* happens to be 0.

If a quadratic equation has no constant—so that it takes the form $ax^2 + bx = 0$—then the *x* can be factored out to get $x(ax + b) = 0$. Then, the solutions are $x = 0$, together with the solutions to $ax + b = 0$. Both factors x *and* $(ax + b)$ can be set equal to zero to solve for *x* because one of those values must be zero for their product to equal zero. For an equation $ab = 0$ to be true, either $a = 0$, or $b = 0$.

A given quadratic equation $x^2 + bx + c$ can be factored into $(x + A)(x + B)$, where $A + B = b$, and $AB = c$. Finding the values of A and B can take time, but such a pair of numbers can be found by guessing and checking. Looking at the positive and negative factors for *c* offers a good starting point.

For example, in $x^2 - 5x + 6$, the factors of 6 are 1, 2, and 3. Now,$(-2)(-3) = 6$, and $-2 - 3 = -5$. In general, however, this may not work, in which case another approach may need to be used.

A quadratic equation of the form $x^2 + 2xb + b^2 = 0$ can be factored into $(x + b)^2 = 0$. Similarly, $x^2 - 2xy + y^2 = 0$ factors into $(x - y)^2 = 0$.

In general, the constant term may not be the right value to be factored this way. A more general method for solving these quadratic equations must then be found. The following two methods will work in any situation.

Completing the Square

The first method is called *completing the square*. The idea here is that in any equation $x^2 + 2xb + c = 0$, something could be added to both sides of the equation to get the left side to look like $x^2 + 2xb + b^2$, meaning it could be factored into $(x + b)^2 = 0$.

Example
$x^2 + 6x - 1 = 0$

The left-hand side could be factored if the constant were equal to 9, , since $x^2 + 6x + 9 = (x + 3)^2$.

To get a constant of 9 on the left, 10 needs to be added to both sides.

That changes the equation to $x^2 + 6x + 9 = 10$.

Factoring the left gives $(x + 3)^2 = 10$.

Then, the square root of both sides can be taken (remembering that this introduces a $\pm$): $x + 3 = \pm\sqrt{10}$.

Finally, 3 is subtracted from both sides to get two solutions: $x = -3 \pm \sqrt{10}$.

The Quadratic Formula

The first method of completing the square can be used in finding the second method, the quadratic formula. It can be used to solve any quadratic equation. This formula may be the longest method for solving quadratic equations and is commonly used as a last resort after other methods are ruled out.

It can be helpful in memorizing the formula to see where it comes from, so here are the steps involved.

The most general form for a quadratic equation is $ax^2 + bx + c = 0$.

First, dividing both sides by a leaves us with $x^2 + \frac{b}{a}x + \frac{c}{a} = 0$.

To complete the square on the left-hand side, c/a can be subtracted on both sides to get $x^2 + \frac{b}{a}x = -\frac{c}{a}$.

$(\frac{b}{2a})^2$ is then added to both sides.

This gives $x^2 + \frac{b}{a}x + (\frac{b}{2a})^2 = (\frac{b}{2a})^2 - \frac{c}{a}$.

The left can now be factored and the right-hand side simplified to give $(x + \frac{b}{2a})^2 = \frac{b^2-4ac}{4a}$.

Taking the square roots gives $x + \frac{b}{2a} = \pm\frac{\sqrt{b^2-4ac}}{2a}$.

Solving for x yields the quadratic formula: $x = \frac{-b\pm\sqrt{b^2-4ac}}{2a}$.

It isn't necessary to remember how to get this formula, but memorizing the formula itself is the goal.

If an equation involves taking a root, then the first step is to move the root to one side of the equation and everything else to the other side. That way, both sides can be raised to the index of the radical in order to remove it, and solving the equation can continue.

Solving Verbal Problems Presented in an Algebraic Context

There is a four-step process in problem-solving that can be used as a guide:

1. Understand the problem and determine the unknown information.
2. Translate the verbal problem into an algebraic equation.
3. Solve the equation by using inverse operations.
4. Check the work and answer the given question.

Example

Three times the sum of a number plus 4 equals the number plus 8. What is the number?

The first step is to determine the unknown, which is the number, or x.

The second step is to translate the problem into the equation, which is $3(x + 4) = x + 8$.

The equation can be solved as follows:

$3x + 12 = x + 8$	Apply the distributive property
$3x = x - 4$	Subtract 12 from both sides of the equation
$2x = -4$	Subtract x from both sides of the equation
$x = -2$	Divide both sides of the equation by 2

The final step is checking the solution. Plugging the value for x back into the equation yields the following problem: $3(-2) + 12 = -2 + 8$. Using the order of operations shows that a true statement is made: $6 = 6$

Geometric Reasoning and Graphing

The four-step process of problem solving can be used with geometric reasoning problems as well. There are many geometric properties and terminology included within geometric reasoning.

For example, the perimeter of a rectangle can be written in the terms of the width, or the width can be written in terms of the length.

<u>Example</u>
The width of a rectangle is 2 centimeters less than the length. If the perimeter of the rectangle is 44 centimeters, then what are the dimensions of the rectangle?

The first step is to determine the unknown, which is in terms of the length, l.

The second step is to translate the problem into the equation using the perimeter of a rectangle, $P = 2l + 2w$. The width is the length minus 2 centimeters. The resulting equation is $2l + 2(l - 2) = 44$. The equation can be solved as follows:

$2l + 2l - 4 = 44$	Apply the distributive property on the left side of the equation
$4l - 4 = 44$	Combine like terms on the left side of the equation
$4l = 48$	Add 4 to both sides of the equation
$l = 12$	Divide both sides of the equation by 4

The length of the rectangle is 12 centimeters. The width is the length minus 2 centimeters, which is 10 centimeters. Checking the answers for length and width forms the following equation: $44 = 2(12) + 2(10)$. The equation can be solved using the order of operations to form a true statement: $44 = 44$.

Equations can also be created from complementary angles (angles that add up to 90°) and supplementary angles (angles that add up to 180°).

<u>Example</u>
Two angles are complementary. If one angle is four times the other angle, what is the measure of each angle?

The first step is to determine the unknown, which is the measure of the angle.

The second step is to translate the problem into the equation using the known statement: the sum of two complementary angles is 90°. The resulting equation is $4x + x = 90$. The equation can be solved as follows:

$5x = 90$	Combine like terms on the left side of the equation
$x = 18$	Divide both sides of the equation by 5

The first angle is 18° and the second angle is 4 times the unknown, which is 4 times 18 or 72°.

Going back to check the answer with the original question, 72 and 18 have a sum of 90, making them complementary angles. Seventy-two degrees is also four times the other angle, 18 degrees.

Translation of Written Phases into Algebraic Expressions

An *algebraic expression* contains one or more operations and one or more variables. To convert written phrases into algebraic expression, there are some key terms to recognize:

- Key terms with addition are *sum, increase, plus, add, more than*, and *total.*
- Key terms with subtraction are *difference, decrease, minus, subtract*, and *less than.*
- Key terms with multiplication are *product, times*, and *multiplied.*
- Key terms with division are *quotient, divided,* and *ratio.*
- Key terms with exponent are *squares, cubed,* and *raised to a power.*

To write a phrase as an algebraic expression, it's necessary to identify the unknown(s) where variables will be used and the words for the correct operation.

<u>Example 1</u>
Write an expression for three times the sum of twice the number *n* plus five.

Three times means *3 x*, twice a number and five means *2n + 5*, and the final expression is *3(2n + 5)*.

<u>Example 2</u>
Write an expression for the total price of $2 per pound for grapes and $3 per pound for strawberries.

The total means the sum. The price for grapes is *2g*, and the price for strawberries is *3s*. The expression is *2g + 3s*.

Practice Questions

1. Which of the following is the result of simplifying the expression:

$$\frac{4a^{-1}b^3}{a^4b^{-2}} * \frac{3a}{b}$$

a. $12a^3b^5$

b. $12\frac{b^4}{a^4}$

c. $\frac{12}{a^4}$

d. $7\frac{b^4}{a}$

2. The graph shows the position of a car over a 10-second time interval. Which of the following is the correct interpretation of the graph for the interval 1 to 3 seconds?

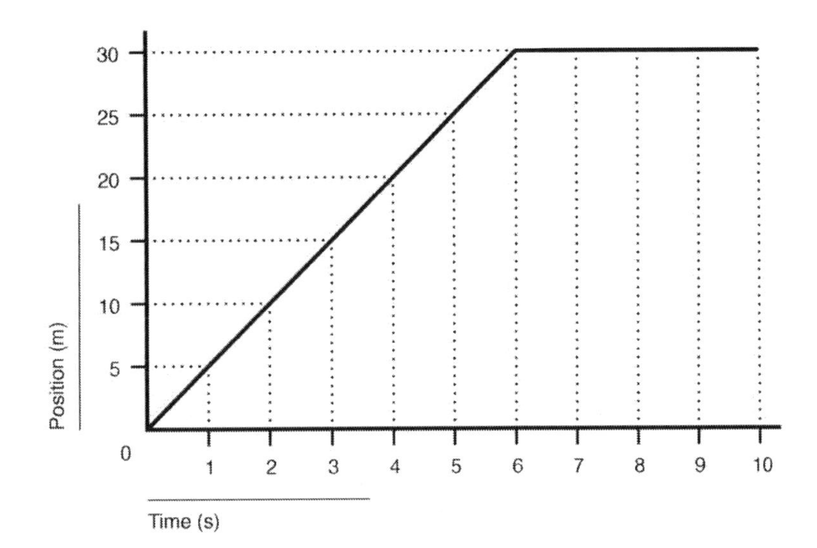

a. The car remains in the same position.
b. The car is traveling at a speed of 5m/s.
c. The car is traveling up a hill.
d. The car is traveling at 5mph.

3. What are the zeros of the function: $f(x) = x^3 + 4x^2 + 4x$?
a. -2
b. 0, -2
c. 2
d. 0, 2

4. The square and circle have the same center. The circle has a radius of r. What is the area of the shaded region?

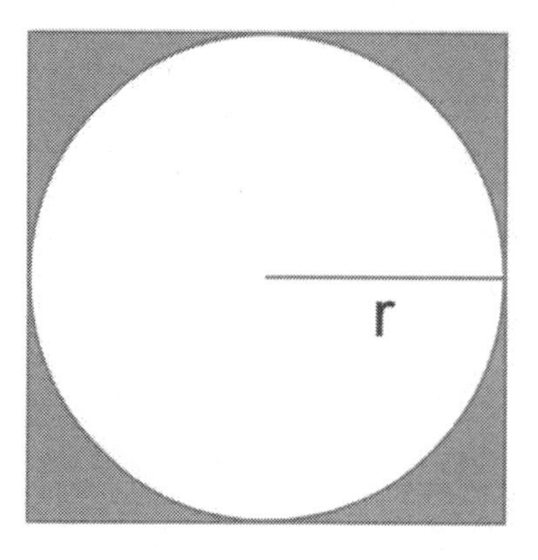

a. $r^2 - \pi r^2$
b. $4r^2 - 2\pi r$
c. $(4 - \pi)r^2$
d. $(\pi - 1)r^2$

5. Which of the following shows the correct result of simplifying the following expression:

$$(7n + 3n^3 + 3) + (8n + 5n^3 + 2n^4)$$

a. $9n^4 + 15n - 2$
b. $2n^4 + 5n^3 + 15n - 2$
c. $9n^4 + 8n^3 + 15n$
d. $2n^4 + 8n^3 + 15n + 3$

6. What is the product of the following expression?

$$(4x - 8)(5x^2 + x + 6)$$

a. $20x^3 - 36x^2 + 16x - 48$
b. $6x^3 - 41x^2 + 12x + 15$
c. $204 + 11x^2 - 37x - 12$
d. $2x^3 - 11x^2 - 32x + 20$

7. How could the following equation be factored to find the zeros?

$$y = x^3 - 3x^2 - 4x$$

a. $0 = x^2(x - 4), x = 0, 4$
b. $0 = 3x(x + 1)(x + 4), x = 0, -1, -4$
c. $0 = x(x + 1)(x + 6), x = 0, -1, -6$
d. $0 = x(x + 1)(x - 4), x = 0, -1, 4$

8. What is the simplified quotient of the following equation?

$$\frac{5x^3}{3x^2y} \div \frac{25}{3y^9}$$

a. $\frac{125x}{9y^{10}}$

b. $\frac{x}{5y^8}$

c. $\frac{5}{xy^8}$

d. $\frac{xy^8}{5}$

9. What is the solution for the following equation?

$$\frac{x^2 + x - 30}{x - 5} = 11$$

a. $x = -6$
b. There is no solution.
c. $x = 16$
d. $x = 5$

10. What is the value of $x^2 - 2xy + 2y^2$ when $x = 2, y = 3$?
a. 8
b. 10
c. 12
d. 14

11. $(2x - 4y)^2 =$
a. $4x^2 - 16xy + 16y^2$
b. $4x^2 - 8xy + 16y^2$
c. $4x^2 - 16xy - 16y^2$
d. $2x^2 - 8xy + 8y^2$

12. If $x > 3$, then $\frac{x^2-6x+9}{x^2-x-6} =$
a. $\frac{x+2}{x-3}$
b. $\frac{x-2}{x-3}$
c. $\frac{x-3}{x+3}$
d. $\frac{x-3}{x+2}$

13. What are the zeros of $f(x) = x^2 + 4$?
a. $x = -4$
b. $x = \pm 2i$
c. $x = \pm 2$
d. $x = \pm 4i$

14. What are the y-intercept(s) for $y = x^2 + 3x - 4$?

a. $y = 1$
b. $y = -4$
c. $y = 3$
d. $y = 4$

15. Is the following function even, odd, neither, or both?

$$y = \frac{1}{2}x^4 + 2x^2 - 6$$

a. Even
b. Odd
c. Neither
d. Both

16. Which equation is not a function?

a. $y = |x|$
b. $y = \sqrt{x}$
c. $x = 3$
d. $y = 4$

17. How could the following function be rewritten to identify the zeros?

$$y = 3x^3 + 3x^2 - 18x$$

a. $y = 3x(x + 3)(x - 2)$
b. $y = x(x - 2)(x + 3)$
c. $y = 3x(x - 3)(x + 2)$
d. $y = (x + 3)(x - 2)$

18. If x is not zero, then $\frac{3}{x} + \frac{5u}{2x} - \frac{u}{4} =$

a. $\frac{12+10u-ux}{4x}$

b. $\frac{3+5u-ux}{x}$

c. $\frac{12x+10u+ux}{4x}$

d. $\frac{12+10u-u}{4x}$

19. What is the answer to $(2 + 2i)(2 - 2i)$?

a. 8
b. $8i$
c. 4
d. $4i$

20. What is the answer to $\frac{2+2i}{2-2i}$?

a. 8
b. 8*i*
c. 2*i*
d. *i*

21. If $4x - 3 = 5$, then $x =$

a. 1
b. 2
c. 3
d. 4

22. Solve for *x*, if $x^2 - 2x - 8 = 0$.

a. $2 \pm \frac{\sqrt{30}}{2}$
b. $2 \pm 4\sqrt{2}$
c. 1 ± 3
d. $4 \pm \sqrt{2}$

23. Which of the following is a factor of both $x^2 + 4x + 4$ and $x^2 - x - 6$?

a. $x - 3$
b. $x + 2$
c. $x - 2$
d. $x + 3$

24. Write the expression for three times the sum of twice a number and one minus 6.

a. $2x + 1 - 6$
b. $3x + 1 - 6$
c. $3(x + 1) - 6$
d. $3(2x + 1) - 6$

25. On Monday, Robert mopped the floor in 4 hours. On Tuesday, he did it in 3 hours. If on Monday, his average rate of mopping was *p* sq. ft. per hour, what was his average rate on Tuesday?

a. $\frac{4}{3}p$ sq. ft. per hour

b. $\frac{3}{4}p$ sq. ft. per hour

c. $\frac{5}{4}p$ sq. ft. per hour

d. $p + 1$ sq. ft. per hour

26. Which of the following inequalities is equivalent to $3 - \frac{1}{2}x \geq 2$?

a. $x \geq 2$
b. $x \leq 2$
c. $x \geq 1$
d. $x \leq 1$

27. For which of the following are $x = 4$ and $x = -4$ solutions?

a. $x^2 + 16 = 0$
b. $x^2 + 4x - 4 = 0$
c. $x^2 - 2x - 2 = 0$
d. $x^2 - 16 = 0$

Answer Explanations

1. B: To simplify the given equation, the first step is to make all exponents positive by moving them to the opposite place in the fraction. This expression becomes:

$$\frac{4b^3b^2}{a^1a^4} * \frac{3a}{b}$$

Then the rules for exponents can be used to simplify. Multiplying the same bases means the exponents can be added. Dividing the same bases means the exponents are subtracted.

2. B: The car is traveling at a speed of five meters per second. On the interval from one to three seconds, the position changes by fifteen meters. By making this change in position over time into a rate, the speed becomes ten meters in two seconds or five meters in one second.

3. B: There are two zeros for the given function. They are $x = 0, -2$. The zeros can be found a number of ways, but this particular equation can be factored into:

$$f(x) = x(x^2 + 4x + 4) = x(x + 2)(x + 2)$$

By setting each factor equal to zero and solving for x, there are two solutions. On a graph, these zeros can be seen where the line crosses the x-axis.

4. C: The area of the shaded region is the area of the square, minus the area of the circle. The area of the circle will be πr^2. The side of the square will be $2r$, so the area of the square will be $4r^2$. Therefore, the difference is $4r^2 - \pi r^2 = (4 - \pi)r^2$.

5. D: The expression is simplified by collecting like terms. Terms with the same variable and exponent are like terms, and their coefficients can be added.

6. A: Finding the product means distributing one polynomial over the other so that each term in the first is multiplied by each term in the second. Then, like terms can be collected. Multiplying the factors yields the expression:

$$20x^3 + 4x^2 + 24x - 40x^2 - 8x - 48$$

Collecting like terms means adding the x^2 terms and adding the x terms. The final answer after simplifying the expression is:

$$20x^3 - 36x^2 + 16x - 48$$

7. D: Finding the zeros for a function by factoring is done by setting the equation equal to zero, then completely factoring. Since there was a common x for each term in the provided equation, that is factored out first. Then the quadratic that is left can be factored into two binomials: $(x + 1)(x - 4)$. Setting each factor equation equal to zero and solving for x yields three zeros.

8. D: Dividing rational expressions follows the same rule as dividing fractions. The division is changed to multiplication, and the reciprocal is found in the second fraction. This turns the expression into:

$$\frac{5x^3}{3x^2} * \frac{3y^9}{25}$$

Multiplying across and simplifying, the final expression is:

$$\frac{xy^8}{5}$$

9. B: The equation can be solved by factoring the numerator into $(x+6)(x-5)$. Since that same factor $(x-5)$ exists on top and bottom, that factor cancels. This leaves the equation $x+6=11$. Solving the equation gives the answer $x=5$. When this value is plugged into the equation, it yields a zero in the denominator of the fraction. Since this is undefined, there is no solution.

10. B: Each instance of x is replaced with a 2, and each instance of y is replaced with a 3 to get $2^2-2\cdot 2\cdot 3+2\cdot 3^2=4-12+18=10$.

11. A: To expand a squared binomial, it's necessary use the *First, Inner, Outer, Last Method.*

$(2x-4y)^2=2x\cdot 2x+2x(-4y)+(-4y)(2x)+(-4y)(-4y)=4x^2-8xy-8xy+16y^2=4x^2-16xy+16y^2$.

12. D: Factor the numerator into $x^2-6x+9=(x-3)^2$, since $-3-3=-6, (-3)(-3)=9$. Factor the denominator into $x^2-x-6=(x-3)(x+2)$, since $-3+2=-1, (-3)(2)=-6$. This means the rational function can be rewritten as $\frac{x^2-6x+9}{x^2-x-6}=\frac{(x-3)^2}{(x-3)(x+2)}$. Using the restriction of x > 3, do not worry about any of these terms being 0, and cancel an $x-3$ from the numerator and the denominator, leaving $\frac{x-3}{x+2}$.

13. B: The zeros of this function can be found by using the quadratic formula:

$$x=\frac{-b\pm\sqrt{b^2-4ac}}{2a}$$

Identifying *a*, *b*, and *c* can be done from the equation as well because it is in standard form. The formula becomes:

$$x=\frac{0\pm\sqrt{0^2-4(1)(4)}}{2(1)}=\frac{\sqrt{-16}}{2}$$

Since there is a negative underneath the radical, the answer is a complex number.

14. B: The y-intercept of an equation is found where the x -value is zero. Plugging zero into the equation for x, the first two terms cancel out, leaving -4.

15. A: The equation is *even* because $f(-x)=f(x)$. Plugging in a negative value will result in the same answer as when plugging in the positive of that same value. The function:

$$f(-2)=\frac{1}{2}(-2)^4+2(-2)^2-6=8+8-6=10$$

yields the same value as:

$$f(2)=\frac{1}{2}(2)^4+2(2)^2-6=8+8-6=10$$

16. C: The equation $x = 3$ is not a function because it does not pass the vertical line test. This test is made from the definition of a function, where each x-value must be mapped to one and only one y-value. This equation is a vertical line, so the x-value of 3 is mapped with an infinite number of y-values.

17. A: The function can be factored to identify the zeros. First, the term $3x$ is factored out to the front because each term contains $3x$. Then, the quadratic is factored into $(x + 3)(x - 2)$.

18. C: The common denominator here will be 4x. Rewrite these fractions as $\frac{3}{x} + \frac{5u}{2x} - \frac{u}{4} = \frac{12}{4x} + \frac{10u}{4x} - \frac{ux}{4x} = \frac{12x+10u+ux}{4x}$.

19. A: This answer is correct because $(2 + 2i)(2 - 2i)$, using the FOIL method is:

$$4 - 4i + 4i - 4i^2 = 8$$

Choice *B* is not the answer because there is no *i* in the final answer, since the *i*'s cancel out in the FOIL. Choice *C*, 4, is not the final answer because we add $4 + 4$ in the end to equal 8. Choice *D*, 4*i*, is not the final answer because there is neither a 4 nor an *i* in the final answer.

20. D: Multiply the top and the bottom by $(2 + 2i)$, the conjugate, to arrive at $\frac{8i}{8}$, which cancels to *i*. Choice *A* is not the answer because the 8's cancel out. Choice *B* is not the answer because the 8's cancel out. Choice *C* is not the answer because 2 is not left, but 8 is.

21. B: Add 3 to both sides to get $4x = 8$. Then divide both sides by 4 to get $x = 2$.

22. C: The numbers needed are those that add to -2 and multiply to -8. The difference between 2 and 4 is 2. Their product is 8, and -4 and 2 will work. Therefore, $x^2 - 2x - 8 = (x - 4)(x + 2)$. The latter has roots 4 and -2 or 1 ± 3.

23. B: To factor $x^2 + 4x + 4$, the numbers needed are those that add to 4 and multiply to 4. Therefore, both numbers must be 2, and the expression factors to $x^2 + 4x + 4 = (x + 2)^2$. Similarly, the expression factors to $x^2 - x - 6 = (x - 3)(x + 2)$, so that they have $x + 2$ in common.

24. D: The expression is three times the sum of twice a number and 1, which is $3(2x + 1)$. Then, 6 is subtracted from this expression.

25. A: Robert accomplished his task on Tuesday in ¾ the time compared to Monday. He must have worked 4/3 as fast.

26. B: To simplify this inequality, subtract 3 from both sides to get $-\frac{1}{2}x \geq -1$. Then, multiply both sides by -2 (remembering this flips the direction of the inequality) to get $x \leq 2$.

27. D: There are two ways to approach this problem. Each value can be substituted into each equation. A can be eliminated, since $4^2 + 16 = 32$. Choice *B* can be eliminated, since $4^2 + 4 \cdot 4 - 4 = 28$. *C* can be eliminated, since $4^2 - 2 \cdot 4 - 2 = 6$. But, plugging in either value into $x^2 - 16$, which gives $(\pm 4)^2 - 16 = 16 - 16 = 0$.

FREE Test Taking Tips DVD Offer

To help us better serve you, we have developed a Test Taking Tips DVD that we would like to give you for FREE. **This DVD covers world-class test taking tips that you can use to be even more successful when you are taking your test.**

All that we ask is that you email us your feedback about your study guide. Please let us know what you thought about it – whether that is good, bad or indifferent.

To get your **FREE Test Taking Tips DVD**, email freedvd@studyguideteam.com with "FREE DVD" in the subject line and the following information in the body of the email:

a. The title of your study guide.

b. Your product rating on a scale of 1-5, with 5 being the highest rating.

c. Your feedback about the study guide. What did you think of it?

d. Your full name and shipping address to send your free DVD.

If you have any questions or concerns, please don't hesitate to contact us at freedvd@studyguideteam.com.

Thanks again!

Made in the USA
Lexington, KY
04 June 2018